THE FOOD POLICE

THE FOOD POLICE

A WELL-FED MANIFESTO ABOUT

THE POLITICS OF YOUR PLATE

JAYSON LUSK

CROWN
FORUM
NEW YORK

CROWN FORUM with colophon is a registered trademark of
Random House, Inc.

Library of Congress Cataloging-in-Publication Data
Lusk, Jayson.
The food police : a well-fed manifesto about the politics
of your plate / Jayson L. Lusk.
p.   cm.
Includes bibliographical references and index.
1. Nutrition—Government policy. I. Title.
RA784.L87   2013
363.19'2—dc23   2012025935

ISBN: 978-0-307-98703-7
eISBN: 978-0-307-98704-4

Printed in the United States of America

JACKET DESIGN BY NUPOOR GORDON
JACKET PHOTOGRAPHS BY FRANK LONGHITANO (FOOD) AND
SPXCHROME/ISTOCK PHOTO (POLICE REPORT)

2 4 6 8 10 9 7 5 3 1

First Edition

For those who wish to eat without a backseat driver

CONTENTS

A SKEPTICAL FOODIE

A catastrophe is looming. Farmers are raping the land and torturing animals. Food is riddled with deadly pesticides, hormones, and foreign DNA. Corporate farms are wallowing in government subsidies. Meatpackers and fast-food restaurants are exploiting workers and tainting the food supply. And Paula Deen has diabetes.

Something must be done.

Or so you would believe if you listened to the hysterics of an emerging elite who claim to know better what we should eat. I call them the food police to be polite, but a more accurate term might be food fascists or food socialists. They are totalitarians when it comes to food, and they seek control over your refrigerator, by governmental regulation when they can or by moralizing and guilt when they can't. They play on fears and prejudices while claiming the high mantle of science and impartial journalism. And their dirty little secret is that they embrace an ideological agenda that seeks to control your dinner table and your wallet.

A chorus of authors, talk show hosts, politicians, and celebrity chefs has emerged as the self-appointed saviors of our food system. They are right about one thing: There is something sinister about our dinner plans. The food police are showing up uninvited.

Whatever tonight's plans, you'd better make room for another guest. It isn't the polite sort who calls ahead and asks what he can bring; it's the impertinent snob who's coming over whether you like it or not, demanding his favorite dish to boot. Be careful what you choose to serve—if isn't made with the finest local ingredients painstakingly bought from small organic farms, expect an evening of condescension and moralizing. And at all costs, avoid the meat from cows fed corn, or you'll have a riot of political correctness on your hands.

Like it or not, the food police will be at dinner. It is impossible to turn on the TV, pick up a book about food, or stroll through the grocery store without hearing a sermon on how to eat. We have been pronounced a nation of sinful eaters, and the food police have made it their mission that we seek contrition for every meal. We are guilty of violating the elite's revelations. Thou shalt not eat at McDonald's, buy eggs from chickens raised in cages, buy tomatoes from Mexico, or feed your infant nonorganic baby food. There can be no lack of faith in the elite's dictates. There are no difficult trade-offs and no gray areas. Thou shalt sacrifice taste for nutrition, convenience for sustainability, and low prices for social justice.

And if we won't willingly repent, the high priests of

politically correct food will regulate us into submission. If you aren't convinced, consider just a few examples of the food police in action.

☞ TRANS FAT BANS

Try a doughnut the next time you're in New York City or Philadelphia. Not as tasty as it used to be, is it? For this we can thank the food police and their war on trans fats.[1] And what a service it has been; we should be grateful to have been taught the shocking truth that too many Oreos and doughnuts are unhealthy!

☞ OUTLAWED HAPPY MEALS

Do you have children? Want to reward them for an A+ on their spelling test? Don't even think about taking them to McDonald's in San Francisco or you'll have a backseat of unhappy campers. The city's board of supervisors tried to ban toys in Happy Meals as a way of "moving forward an agenda of food justice."[2] The hypocrisy is astounding. So now, "in the City by the Bay, if you want to roller skate naked down Castro Street wearing a phallic-symbol hat and snorting an eight-ball off a transgender hooker's chest while underage kids run behind you handing out free heroin needles, condoms and coupons . . . *that's* your right as a free citizen of the United States. But if you want to put a Buzz Lightyear toy in the same box with a hamburger and fries and sell it, you're outta line, mister!"[3]

☞ TWINKIE TAXES

Despite the economic research clearly showing that fat taxes will do little to slim our waistlines, food police across the nation are hiking food prices by implementing various forms of taxes on soda, fat, and fast food. In his book *The World Is Fat,* University of North Carolina professor Barry Popkin says, "Taxing the added sugar in beverages is a favorite strategy of mine."[4] So now you know whom to thank when your grocery bill is 100 percent higher and half as tasty.

☞ LOCAL FOOD SUBSIDIES AND PURCHASING REQUIREMENTS

Rather than using our tax dollars to shore up Medicare or Social Security, the food police want to subsidize your neighbor's purchase of local asparagus. With the help of lawmakers, bestselling author Michael Pollan wants to "require that a certain percentage of that school-lunch fund in every school district has to be spent within 100 miles."[5] Tough luck for citrus-loving children in Minnesota.

☞ AFFIRMATIVE ACTION FOR COWS

The Obama administration tried to implement rules on the "fair" pricing of livestock that would have required ranchers and meatpackers to justify to the government the prices they freely pay and accept for cattle. Cowboys across the nation would have had to tell Uncle Sam why they paid more for a hearty registered Angus than a scrawny half-breed. In a move projected to have cost farmers and consumers more than $1.5 billion annually, and radically altered the structure of the livestock sector, the food police unwittingly married George Orwell's two greatest works by bringing Big Brother onto Animal Farm.[6]

☞ DIRT TAXES

As if just now realizing that corn grows in the ground, the Environmental Protection Agency is trying to implement rules enabling it to fine farmers if their tractors kick up too much dust. Charlie Brown's friend Pig-Pen had better watch his back.

☞ **FRUIT AND VEGGIE SUBSIDIES**

Convinced that the food industry is "incapable of marketing healthier foods," *New York Times* food writer Mark Bittman wants to expand the welfare state by cajoling farmers and consumers into growing and buying the stuff he prefers that we eat. Bittman wants to enact policies that will "subsidize the purchase of staple foods like seasonal greens, vegetables, whole grains, dried legumes and fruit."[7]

☞ **FOIE GRAS BANS**

Only after a groundswell of outrage from chefs and restaurant-goers did the city of Chicago rescind its ban on the duck and goose liver delicacy. Undeterred by the suffering palates of their Midwestern brethren, the state of California now bars its own citizens from buying what can be found on almost every street in Paris.

☞ **REGULATING HAPPY HENS AND HOGS**

Despite the fact that fewer than 5 percent of consumers buy cage-free eggs or pork, activist efforts have led to ballot initiatives and legislation in at least eight states requiring farmers to use cage-free production systems. The laws force farmers to adopt practices that most consumers aren't fully willing to pay for.[8]

☞ **TECHNOPHOBIA**

Wirelessly pontificating on Facebook and Twitter via their 64 GB iPads, the food police welcome new technology—that is, unless it relates to food. Unwilling to accept the scientific evidence on the potential price-reducing and safety-enhancing advantages of new food technologies, the food police promote bans, taxes, restrictions, and propaganda on technologies such as preservatives, irradiation, biotechnology, cloning, and pesticides. They even spread fear about age-old practices such as pasteurization, hybrid plant breeding, and grain-fed livestock.

☞ SCHOOLROOM INDOCTRINATION

The food police seek to indoctrinate our children, not by teaching food science and nutrition, but by advancing the cause of fashionable foods. Famed food activist and restaurant owner Alice Waters wants "a total dispensation from the president of the United States who will say, 'We need a curriculum in the public school system that teaches our kids, from the time they are very little, about food and where it comes from. And we want to buy food from local people in every community to rebuild the agriculture.'" She says that we "*must* get Obama to understand the pleasures of the table."[9] His wife was listening. Michelle Obama's signature childhood nutrition bill took $4.5 billion away from food stamp recipients to expand the federal government's role in regulating school lunches and significantly increased food costs to local schools throughout the nation.[10]

☞ RESTRICTIONS ON FREEDOM OF SPEECH

A team of four government agencies—the Federal Trade Commission, the Food and Drug Administration, the Department of Agriculture, and the Centers for Disease Control and Prevention—have banded together under the auspices of an Interagency Working Group to recommend prohibitions against certain food advertisements to children. Some like-minded members of the food elite want to invoke a version of the Fairness Doctrine, demanding equal airtime to run government and activist-sponsored ads about food.

These are but a few examples of the growing intrusions by the food police. Taken in isolation, any one of the regulations might not seem so bad and may even appear helpful. Therein lies the danger. You won't yet find a single omnibus piece of legislation restricting your food freedoms, but the planks in the road are slowly being replaced without the travelers even realizing construction is under way. And guess who is left to pay the toll at road's end? With a ban on trans fats here, a fat tax there, here a local foods subsidy, there a pesticide ban, everywhere an organic food—before you know it, Old McDonald has a new farm.

Michelle Obama's White House garden was a symbolic nod granted to the growing reality of a movement that seeks more control over what we eat. Even New York City mayor Michael Bloomberg supports the encroaching government hand in food. After congratulating himself on helping New York City ban trans fats, successfully pressuring food companies to reduce salt, and selectively licensing "green" produce vendors, Bloomberg told the UN General Assembly that "Governments at all levels must make healthy solutions the default social option. That is ultimately government's highest duty."[11] Never mind national defense or the government's duty to life and liberty or leaving unenumerated powers to the states: healthy food is now apparently the be-all and end-all of good government.

These are many of the same people who scream, "It's a woman's body," any time the subject of abortion comes up. According to them, a woman has the right to do what she wants with her body. That is, unless she wants to eat something deemed morally defunct. You know, something as reprehensible as a Nestlé Toll House cookie or a Big Mac.

Susan Dentzer, former PBS news correspondent and current editor in chief of *Health Affairs,* wants to change our eating habits by creating a "broadbased set of interventions comparable in scope to the four-decade assault on smoking."[12] Others are "turning to multilevel interventions, which . . . target . . . the individual, the social network, the community, and policy."[13]

What exactly are these interventions? Consider the proposal mentioned in *Harvard Magazine*: "There was once a very successful U.S. government program aimed at changing eating habits . . . It happened during World War II, and it was called 'food rationing.' They made it a patriotic thing to change the way you ate. The government hired the best people on Madison Avenue to come to Washington and work for the War Department. It worked splendidly."[14]

Food rationing! Really? I suspect that many of the millions of people who actually lived through food rationing during World War II would recall the policy as being anything but splendid. But apparently if the food police are to have their way, we must be prepared to stand in line and register for a coupon booklet to buy sugar, bread, and flour. It's as if we have already forgotten the dreaded food lines of the USSR.

⊘

While these obsessions about food have captivated a modern generation of consumers far removed from the farm, I must admit to being befuddled. The sirens of paranoia and pronouncements of cataclysmic catastrophes ring hollow. Although many of the food police are truly concerned about our diets and health, and not all are avowed ideologues, the real trouble comes when we take a closer look at the consequences of the proposed solutions offered by those with a romanticized vision of agriculture. If the food elite are disenchanted with the way food is now produced, many in the agricultural community are mystified that city-dwelling journalists presume to know so much about how to run a farm. It might be fun to play farmer for a few days, but a lifetime of such work is not only difficult for most to imagine, but something that few truly desire. The truth is that for the past one hundred years, people have been leaving the countryside in droves seeking a better life in town. I know, because I'm one of them.

I know the people who run the farms and factories that are demonized by the food police. You've been shown but one small part of the picture. Bestselling authors and journalists tell the stories of the folks selling a few chickens at the local farmers' market, but where are the people who actually feed America?

My love of food began as a child, when I entered my mother's beef Stroganoff recipe in a local 4-H cooking competition. Although the judges weren't particularly impressed with my efforts, I became fascinated with the way disparate food ingredients could be combined to produce an entirely new and wonderful taste. Perhaps it was my parents' insistence that we children eat at least three bites of everything put on our plates that led me to see food as a medium of experimentation and learning. And learn I did. In high school, I became one of the best dairy food judges in the state of Texas. (Yes, there is such a thing as dairy food judging, and yes, Future Farmers of America [FFA] students all over America still do it today.) As strange as it may seem, being able to discriminate among thirty different milks and cheeses was apparently sufficient qualification for receiving a scholarship to study food technology in college. Not only did I develop a palate, I learned where food comes from.

There weren't many options for out-of-work teenagers in rural West Texas, so every summer from 1987 to 1994, I donned boots and a baseball cap to rid the earth of the weeds that were the bane of the local farmers' existence. Those years spent working cotton and soybean fields and waking up mornings to feed sheep and hogs taught me an important lesson: farming is not the romanticized profession it is often made out to be. It is a lot of hot, sweaty, backbreaking work. Real pigs that live on farms aren't anything like Wilbur or Babe; they bite, they poop, they break fences, they stink, and they can't go a day without food or water, no matter what the outside temperature is or what vacation plans have been made.

Food journalists like to talk about being one with the land and of "nature's logic,"[15] but make no mistake about it, agriculture, by its very definition, is a struggle against nature. After all, there's nothing more natural than death. Nature isn't our friend. It's trying to outcompete us.

Those summers working in the fields were all the motivation I needed to finish college. If farming wasn't for me, I thought my place in life might be one step up the food chain.

I decided to study food science and technology at college. There I learned more valuable lessons. Eating is inherently risky, and it is the development of food technologies (some very old and some very new) that make eating safe. But perhaps the best lessons I learned in college were during the summers working in a food processing plant near Dallas.

There I discovered that the people working in agri-businesses—both the owners in the corner offices and the factory-line workers screwing caps onto salsa jars—were regular people just like you and me. Neither was exploiting or being exploited, despite what I'd been taught in my history classes or read in books from ivory tower academics (of which, ironically, I am now one). Even though it was hard work, almost everyone I encountered in that processing plant was committed to putting out high-quality, safe food, probably because they knew that if they didn't, their competitors would. At my first day on the job, I received a tongue-lashing when I unwittingly cleaned the inside of a stainless-steel mixing tank with a brush designated for floor sweeping. The corrective came not from a manager but from a minimum-wage-earning high school dropout, who simply said, "How'd you like to stick your mouth on the ground? 'Cause that's what you're doing." After all, the food that came out of that tank was the same food he fed his own family.

My mistakes didn't end there. Once, while working a late-night shift, I was responsible for measuring the thousands of pounds of ingredients to make a fettuccini sauce for a major restaurant chain. The next day, much to my horror, I learned that I had included twice the amount of butter required for the sauce. (Despite three semesters of engineering calculus, I had mistaken kilograms for pounds!) Having lost the company thousands of dollars, I could easily have been fired, but I wasn't. This experience, and others like it, helped me realize that it was the interaction between food *and people* that was most interesting, and there was no better way to look into the matter than through the lens of economics. My passion became learning why people ate the things they did and figuring out how people dealt with the difficult trade-offs between health, safety, and taste when their paychecks wouldn't let them have it all. I became a food economist.

When people find out I'm an economist, they usually ask me what will happen to stock prices or interest rates. I haven't the faintest idea what to tell them, and truth be told, most economists don't know either. Deep down, economics isn't really about financial markets. It is the study of how we make decisions when our wants are bigger than our wallets. Food economists spend their time thinking about how to eat more and better with less; providing information about food supply-and-demand conditions to farmers, agribusinesses, nonprofits, consumers, and government agencies; and projecting the impacts of private and government decisions.

I have spent the last fifteen years studying the economics of food—trying to figure out things such as how much consumers are willing to pay for better-tasting meat and why they seek to avoid controversial food technologies. I've written hundreds of academic papers on food economics, consulted nonprofits, agribusinesses, and governments, served on the editorial councils of a half dozen of the top academic journals in agricultural and environmental economics, and sat on executive boards of the three largest U.S. agricultural economics associations. I tried to approach the study of food regulation from an objective standpoint by comparing the costs and benefits of the policies in question—seeing which actions and policies made the best use of our scarce resources given all our competing desires. I labored under the assumption that this was the key issue in determining the merits of a regulation. I was naive.

Today, food policies are increasingly motivated by ideology rather than projected economic consequences. The merits of a policy are primarily judged not by whether they benefit all the people affected but by whether they advance the fashionable agenda of a new food elite. It was once the case that those proposing a new food policy had to offer serious analysis illustrating its impacts, but now the simple mention of a food problem—the rise in obesity or poor nutritional intake among the poor—is apparently sufficient justification for a policy intervention regardless of any evidence that it might work. All that is required is to demonstrate sufficient empathy for those in need. Jonah Goldberg recognized that modern liberalism "is an ideology of good intentions" wrapped up in a "cult of action."[16] And so it is with the food police.

A few years back I was asked to join a small group of faculty members in a discussion with a Burger King executive who had been invited to give a lecture on campus. I was eager to hear about the company's challenges and business strategies, but one of my then-colleagues, a professor from the nutrition department, apparently saw fast-food restaurants as one of the greatest evils ever conceived. She lectured the executive on the sodium and fat content in the Whopper and, with all the righteous indignation she could muster, asked how he felt about being responsible for the large number of heart attacks that occurred each year in America.

The executive's response was as revealing as the professor's diatribe. He rattled off the sales statistics for Burger King's salads and other healthy offerings; those items had been an abysmal failure. Burger King's primary customers, eighteen-to twenty-five-year-old males, apparently don't order salads, and the executive stressed that if Burger King wanted to remain in existence and continue to write paychecks for its more than thirty-five thousand employees, it had no choice but to offer what its customers were willing to buy. The nutrition professor was undeterred by the reality of what people actually wanted in the real world. She and others like her have an agenda that must be pursued regardless of whether the intended beneficiaries actually want it.

I finally realized the extent of the absurdity last year, when I was asked to speak to the teachers of my university's freshman English composition courses. I walked across campus prepared to give a five-minute spiel on writing strategies. Expecting questions on how I prepare students for the rigors of academic publishing, I was amused when a graduate student shot up her hand and asked what I thought about the documentary *Food, Inc.*

For the next half hour, I engaged in a lively discussion with about seventy-five graduate students and professors of English. I wasn't surprised that they had questions about food and agriculture, and I was happy to answer them. What surprised me was the absolute moral certitude permeating the air. Many in the room had no doubt they were being poisoned and fattened by an out-of-control food system. What facts did these folks have to bolster their case? Nothing more than what was presented in *Food, Inc.,* along with a few innuendos picked up in the Sunday paper. I would never have dared question them on the finer points of Shakespeare or Dostoyevsky. Yet they were certain that my explanations for why food is produced the way it is were wrong.

More shocking still was the fact that this debate took place not in Berkeley, California, or Cambridge, Massachusetts— there I might have expected some unusual views about agriculture, if for no other reason than that these locations are so far away from where most of our food is made. But this was Stillwater, Oklahoma. All these people had to do to see large-scale production agriculture was look out the window. Most of them probably lived within a mile of a real-life, flesh-and-blood farmer. They had no doubt stood next to a Monsanto chemical salesman in the checkout line at the grocery store. Yet they had come to believe that farmers were destroying their land and that agribusinesses were manipulating farmers and polluting the food supply. Many of my colleagues had swallowed a story whole from the food elite when they could have gotten closer to the truth simply by talking to their neighbors.

Unlike the food police, I do not have an agenda to change what you eat. Nor am I a defender of the status quo. One of the things I most like about being an economist is that we are generally agnostic about people's preferences: Animal-loving vegans and BBQ-licking carnivores are treated all the same. Human nature is recognized for what it is. Rather than chastising people for who they are, a more fruitful path forward is to look at the outcomes from our interactions with others and see what results are produced by our different rules, norms, markets, and policies. The famed Austrian economist Ludwig von Mises perhaps said it best: "Economics . . . abstains from any judgment of value. It is not its task to tell people what ends they should aim at . . . Science never tells a man how he should act; it merely shows how a man must act if he wants to attain definite ends."[17]

The food police wish people were different and seek to change our preferences and behavior through coercion and propaganda if necessary. Of course, reasonable people can change their minds about the desirability of modern food production, but ideally these decisions would be made with accurate knowledge of the world surrounding them. I wrote this book because I can no longer sit back while the truth about food, and economics, is so badly distorted.

Muckraking journalists are right to point out the errors and limitations in our modern food system, but we shouldn't let that blind us to all that is good about how we eat. We have forgotten what we do well. As a result, we are on the verge of accepting cures that are far worse than the disease. It is fashionable to belabor all that is wrong with food and then pitch grand schemes that promise to save us all. I intend to do just the opposite. I seek to remind readers that we live in the greatest food-producing and -consuming nation in the history of the world, and to provide cautionary tales of the impotency, and sometimes outright lunacy, of the policies sought by the food police.

Lest one believes my claims are merely rhetorical embellishments, let's consider some facts about the state of food in America.

☞ MORE LEISURE TIME

Cooking and cleaning have gotten much easier and more convenient. Today we spend 40 percent less time in food preparation and 81 percent less time on meal cleanup than we did in the 1960s.[18] If that change doesn't sound good to you, go talk to your grandmother.

☞ MORE MONEY IN YOUR POCKET

Food is more affordable in the United States than any-where else in the world. Americans spend half as much of their income on food prepared at home as the Germans, French, or Italians, and about a third as much as our Mexican neighbors.[19] Food is also much more affordable than it once was. As a portion of our income, we currently spend about half as much on food as compared with the 1950s and about one quarter less than we did in the 1970s.[20] Today we spend less than 10 percent of our income on food, which leaves 90 percent for all the other things in life we enjoy, such as art, baseball, Disneyland, iPhones, and computers.

☞ THE LUXURY OF EATING OUT

We can now afford to eat out and let others cook and clean for us. Eating out was a rare treat when I was a child; today my kids expect to be waited on several times a week. In the 1970s, only about 28 percent of Americans' food budget was spent eating out; today Americans spend about 43 percent of their food budget away from home.[21] There has been a more than 60 percent increase in the number of restaurants per person since the early 1970s. The ability to grab a quick meal and let someone else clean up is one of the many factors that has allowed more women to pursue a career outside the home.[22]

☞ MORE FRUITS AND VEGETABLES

We are eating many more fruits and vegetables. Per capita consumption of vegetables has increased by 20 percent since the 1970s; consumption of fresh vegetables increased 30 percent over this period.[23] Similarly, per capita consumption of fruit has increased by 10 percent since the 1970s; consumption of fresh fruit increased 20 percent over this period.[24]

☞ SAFER FOOD

Our food system is among the safest in the world and is getting safer. Laboratory-confirmed infections from the most common food-borne bacteria have fallen about 25 percent since the mid-1990s. Infections from specific dangers such as *E. coli* 0157 have fallen as much as 41 percent over that time.[25]

☞ LIVING LONGER

In 1900, average life expectancy in the United States was only 47.3 years. In 1970 it was 70.8. A baby born today can expect to live more than 78.7 years.[26] We live longer, in part, because of better medical technologies, but also because we have access to a more abundant, diverse, and nutritious food supply.

☞ MORE PRODUCTIVE FARMERS

Science and innovation have allowed farmers to reap much more food from the same amount of land. From 1900 to 2010 the average amount of corn produced per acre more than quadrupled, increasing a whopping 440 percent! Wheat yields increased 280 percent. Even yields for obscure commodities such as flaxseed have increased over 270 percent since 1900. In the last twenty years alone, corn yields increased 30 percent, wheat yields increased 17 percent, and soybean yields increased 28 percent.[27]

☞ FEWER AND SAFER INSECTICIDES

It doesn't matter how farmers grow crops; there will always be worms, beetles, and other pests that want to eat the food before we do. Fortunately, farmers have found better ways to stamp out the competition. Crop farmers today use about half as much insecticide as they did in the 1960s.[28] Thanks to biotechnology, insecticide use in corn fell 80 percent from 1997 to 2007. Moreover, USDA and EPA researchers studying agricultural chemical use concluded that "[i]n recent years, we have witnessed a significant trend toward replacing relatively hazardous active ingredients with less hazardous ones."[29]

☞ FINANCIALLY SECURE FARMS

Even after accounting for inflation, farmers earned 3 percent higher net incomes in the 2000s than they did in the 1990s. The value of their farm assets also rose, by almost 30 percent in real terms during that time.[30] In fact, farming is no longer the profession of paupers. Although the income of farmers lagged behind that of nonfarm households for almost a century prior to the 1990s, the median income of farm households has exceeded the median income of all U.S. households in every year from 2000 to 2010. Median farm income averaged 10 percent higher than nonfarm household income over that same period.[31]

☞ MORE CHOICE

A hundred years ago, many households were constrained to eat only what they could grow. We now have the opportunity to find flavors and textures particularly suited to our own palates. New product introductions by the food and beverage industry have increased almost 100 percent in the past two decades alone.[32]

Don't take my word for it. Even the U.S. Department of Agriculture, in 2003, concluded, "There is no doubt that [the food system] delivers—more nutritious food with wider variety; improved safety, with less environmental impacts; and greater convenience than at any time in the Nation's history."[33]

I'm not saying all food trends are heading in the right direction. Rather, my claim is that the actions of the food police have produced a misleading picture of food and agriculture. The food police rely on a distorted view of reality to gain support for a sweeping agenda that, if successful, will cause more harm than good.

In the following pages, I endeavor to convey the enormity of what is at stake. I trace the history of the food police and uncover their origins in the thoughts and attitudes of last century's Progressives. Today's food police, sympathetic to the anticapitalist leanings of the socialists of yesteryear, have a new weapon in their arsenal: behavioral economics. I show that behavioral economics is the intellectual engine behind the food elite's justifications for overriding our preferences and decisions with their own. Given this general backdrop, I then take specific aim at the fetishes of the food police (organic and local food) and their fear of biotechnology, all of which is underpinned by an untenable ideology of food. Along the way, we'll see that the food police's favored policies—from fat taxes to veggie subsidies to calorie labels at restaurants to strategic grain reserves—will hurt the poor, damage the environment, limit choice and innovation, and ultimately do little to alleviate their foodie angst.

The food police don't have a monopoly on passion for food. I love to eat, to cook, and to visit the farms and factories that make it all possible. I've spent a lifetime learning where food comes from and how to make it better. I believe our freedoms are worth fighting for—from what farmers choose to do on their own land to what we decide to stick on our own forks. My desire is that farmers and agribusinesses have the freedom to compete for a place on your dinner plate, and that you have the knowledge to eat, with a clear conscience, what you wish.

THE PRICE OF PIETY

Tune in to the *Martha Stewart Show* or the Food Network, watch *Food, Inc.* or *Supersize Me,* or pick up a book such as *Fast Food Nation* or *The Omnivore's Dilemma,* and you'll find that there is an emerging consensus about food. What we hear isn't good news. In fact, it's downright frightening. Yale professor Kelly Brownell tells us that we live in a "toxic food environment."[1] Bestselling author Michael Pollan says that "Americans have a national eating disorder."[2] After showing a picture of an atomic explosion, *New York Times* columnist and food writer Mark Bittman proclaimed that our modern food production system is leading to "a holocaust of a different kind."[3]

The prescription for our ailments is local, organic, slow, natural, and unprocessed food along with a healthy dose of new food taxes, subsidies, and regulation. If we would only repent our wicked ways and shackle our oppressors, we'd reap better-tasting, healthier, sustainable, and equitable food. To save the planet and our very lives, we must return to nature and abandon modern agriculture. These are the sacred cows of a growing food movement. Yet sacred cows make the tastiest burgers.

Is it possible that what you've read about modern food is all wrong? We humans are notoriously bad at judging small risks. It's also human nature to focus on the negative and the emotional. The result is that stories claiming that we are drowning in a river of cheap, pesticide-riddled corn are more salable in the media than reports showing the benefits of lower prices at the supermarket. We fret over the dangers of food pesticides when we're 1,600 times more likely to die in a car accident and even 15 times more likely to die from drowning in the bathtub.[4] When activist organizations and government agencies require crises to secure donations and public funding, there is a natural incentive to slant the facts. In fact, our vision of reality can become badly distorted.[5]

Did you know, for example, that science is now beginning to suggest that despite all we've been told, cutting salt intake has little effect on the chances of having a heart attack or stroke and may even increase the chance of death?[6] Yet with every new study questioning the effectiveness of salt-reducing diets, the food police ramp up their efforts to force food processors and restaurants to cut back on sodium. It is time to rethink the rethinking of the modern American food system.

We can deliberate any food policy, but something larger than costs and benefits is at play. As John Donvan of ABC

News put it when moderating a debate on organic food, "What strikes me about this debate is that the tone here is as bitterly partisan as anything that's happening in Washington. And I'm curious about why that is. And it's on both sides. It's also from all of us here in the hall. There is a nasty feeling to this issue."[7]

The debate is nasty because our freedom is at stake. On one side are farmers who want to work, and consumers who want to eat, as they please; on the other side are the self-proclaimed saviors of the food system, who want to make decisions for us. The food elite have appointed themselves our caretakers. They seek the power to steer food production and choice, claiming to know better than farmers and consumers. It is time we regained control of our forks and farms, and rightfully assumed responsibility for our own health, environment, and pocketbooks.

The surprise for the average grocery shopper, not to mention many fashionable foodies, is that the food zeitgeist did not result from careful thought and scientific investigation; it is the result of a concerted agenda to remake our food system. Avid buyers of organic who are just trying to protect their families and the environment are largely unaware of the undercurrent driving the modern food movement, but the food elite understand it all too well.

Even a casual reading of the bestselling books on food and agriculture reveals an explicit disdain for the system of free enterprise. Michael Pollan writes of "another example of the cultural contradictions of capitalism—the tendency over time for the economic impulse to erode the moral underpinnings of society."[8] Pollan sympathetically portrays a farmer who claims, "The free market has never worked in agriculture and it never will."[9] Stan Cox, a scientist at the Land Institute, opines about the state of modern agriculture when asserting, "If we find no alternative to capitalism, the Earth cannot be saved."[10] Author and nutritionist Marion Nestle is "increasingly convinced that many of the nutritional problems of Americans . . . can be traced to . . . a highly competitive marketplace."[11] The American Public Health Association, a leading proponent of new dietary regulations, even has a Socialist Caucus. Food activist Frances Moore Lappé reveals that the modern food movement "is at heart revolutionary."[12] In fact, the modern slow food movement grew from "intellectuals on the political left who were . . . supporters of communist ideology."[13]

The food elite have found the problem underlying our modern food dilemmas, and it is nothing short of capitalism and individual freedom. They demand a revolution in food—a dictatorship of the foodie proletariat.

Of course I'm not saying that everyone who wants healthier or tastier food is a socialist. Far from it. The question isn't whether we want healthier and tastier food. Who doesn't want that? The question is *how* to get it and what it *costs*. What the food police offer is not a realistic plan for food but an ideology. They offer compassion, but it's easy to be caring and generous with other people's property and choices. The modern food movement did not arise in a vacuum, and the food elite providing the intellectual underpinnings are far from dispassionate observers. Here's a recipe for how they work.

MAMA'S HOMEMADE FOOD CRISIS

---

1 cup feigning compassion

1 28-oz. can paternalism in moralizing syrup

3 misleading interviews with farmers picked from the
　　subscription list of *Mother Jones*

1 tablespoon inequality aversion harvested from a
　　Wall Street Occupier

1 stick butter (margarine will work if you're anti-cow)

2 eggs laid by chickens on Rhodes scholarships

1 jar elitism siphoned through the Ivy League

3 cups antibusiness ethos

1 dash hypocrisy

STEP 1: Mix first four ingredients (for best results, none should be local), taking care to remove any inconvenient facts that bubble up. **STEP 2:** Blend in butter and eggs (only local products here, to avoid hint of impropriety) and work until a smooth river of mainstream media and government funding emerges. **STEP 3:** Whip batter into media frenzy and an apparent epidemic is produced. **STEP 4:** Add in the elitism and antibusiness ethos and place in pressure cooker until Congress passes new legislation. **STEP 5:** If desired legislation is not obtained, add last ingredient, bypass the legislative process, and use the president's executive order and bureaucratic fiat to complete the recipe. Feel free to rig the rules to get corporate photo ops to support the cause. **STEP 6:** Quickly move on to next problem before unintended consequences appear.

⊘

The police can be an important source of security and support for a neighborhood. A responsible police force works at
the behest of the citizens, with their explicit direction and
consent. But when the police believe that the citizenry are no
longer capable of knowing what is in their own best interest
and when they use their power to advance their own ideology, good-bye Andy Griffith and enter the Gestapo. We do not
yet have martial food law, but no totalitarian state in history
emerged without the stage first being set with ideas and the
populace giving up incremental powers in the name of the
greater good. It is against this danger we must be on guard. It
was the philosopher David Hume who said, "It is seldom that
liberty of any kind is lost all at once."[14]

The truth is that the food police don't seem to understand
the challenges faced by everyday food consumers. Celebrity
chefs are accustomed to serving dinners to people with fat
wallets. Nutritionists are trained to evaluate every morsel
that goes into our mouths. Out in the real world, you and I
have real trade-offs to make. We can eat the fifty-dollar
roasted squab or we can make our mortgage payments; we
can cut out the saturated fats and sugars or we can eat something tasty.

In the idealized world of the food police, we'd all eat in
fine restaurants—of course, passing on the foie gras appetizer
if the nutritionist or animal rights activist is watching from
an adjacent table or if you live in one of the places, such as
California, where the food police have already banned it—and
someone else would pick up the bill. We'd eat leafy greens and

feel just as satisfied as if we'd eaten prime rib. This is nothing more than utopian nonsense.

F. A. Hayek, Nobel Prize winner in Economics, adequately summed up the difficulty some people have acknowledging the reality of tough choices: "People just wish that the choice should not be necessary at all. And they are only too ready to believe that the choice is not really necessary, that it is imposed upon them merely by the particular economic system under which we live. What they resent is, in truth, that there is an economic problem."[5] Wishing we didn't have to make tough economic trade-offs is one thing. Ignoring them altogether is downright dangerous.

By all means, if you want advice on finding delicious food, look up a Michelin-rated chef. Want to drop a few pounds? Find the advice of a trained nutritionist. (Which one, I'm not exactly sure—just take your pick from the hundreds of diet books on the market. You can lose weight by eating only meat. Or is it no meat? Or is it only liquids? Or is it only by thinking happy thoughts? Just save the fifteen bucks, eat less and exercise more, and that should do the trick.) If you want to know how to eat eggs from happy chickens, go find popular books by philosophers. (Better yet, go for the more obscure books by the animal scientists.) But don't ask these folks to run your life.

That's the problem with specialists. They are . . . well, special. They take the *one* thing that is important to them and assume it should be most important to everyone else, too. Anthony Bourdain, the bestselling author and former chef, thinks everyone should know how to cook. He asks of his idealized world, "What should *every* man, woman, and teenager know how to do?"[16] According to Bourdain, we *should* all know, at a minimum, how to chop an onion, make an omelet, prepare a vinaigrette, cook meat and veggies to the right degree of doneness (without a thermometer no less), clean a fish, steam a lobster, and make beef stock from scratch. It's this kind of thinking that leads folks such as Alice Waters to promote policies to turn public schools into training grounds for future farmer-chefs.

I can make vinaigrette, but I can't imagine why *everyone* should or why our world would be better if only Suzie in Omaha knew how to make beef bourguignon. After all, Newman's Own is pretty tasty. Pre-chopped onions and premade beef stock are all readily available in my supermarket—and they're cheap, too. If I want steamed lobster or well-cleaned fish, I know which restaurant to patronize.

As if we didn't have better or more enjoyable things to do with our time than learn to cook rice by the pilaf method. What makes our modern world so wealthy is the diversity of our interests and our willingness to trade what we do well for what others do well. Why would a *chef* want you to know how to cook well, thus rendering his services obsolete? It must be a natural human tendency to take those values from our own narrow world, so all-encompassing to us, and project them onto others who haven't a clue about the Culinary Institute of America or what its graduates such as Bourdain want. I'll tell you what I want from the Culinary Institute of America: I want it to produce graduates I'm willing to pay to do all the things they think I *should* know how to do.

At least Bourdain limits his idealized vision for you and me to the pages of a wildly entertaining book. Other foodies aren't so restrained. The real trouble comes from those who seek the power to make decisions for us. By now you won't be surprised to hear that nutritionists think we should all be eating more nutritiously, environmentalists think we should all be eating in a way that is more environmentally friendly, and social justice advocates think the poor should have more access to veggies.

Herein lies the danger in our "food crisis." An "epidemic" such as obesity provides the traction for the specialists to implement their vision for the world. Everything unseemly is said to cause obesity: supporting corporate farms and manipulative agribusinesses, being poor, eating processed foods, eating too much meat, eating nonlocally, and on and on. With one problem we unite an unholy alliance of the antibusiness activist, the social justice activist, the chef, the nutritionist, the animal rights activist, the environmentalist, and all the rest. Never mind the fact that none of these people would agree among themselves how to make our brave new world. But that's beside the point. They have a problem that can advance the agenda, regardless of whether they know where we are advancing. The reality is that we each have our own individual goals and values, and there is no single objective free of tough trade-offs upon which we can all agree.

Obesity is but one item in the smorgasbord of food crises. After all, "[y]ou never want a serious crisis to go to waste"— especially if it's one you've helped create.[17] The overarching solution sought by the alliance of the food police is a return to nature. The food elite seek to achieve a kind of a false traditionalism in food that never actually existed and cannot work in practice. The story of the modern progressive food movement is largely one about a reaction to technological change in food and agriculture. Ideology has led the food police to denounce the benefits of technology as accruing to greedy agribusinesses, fast-food restaurants, and corporate farms, and has led to an agenda seeking false agricultural idealism. By denying the massive efficiencies of modern food technologies, the food police are starving the folks they say they're trying to help. In their vexation over obesity, many of the food police seem to have forgotten that more than 14 percent of American households—more than 17 million people—are currently classified as "food insecure," which means they have financial difficulties providing enough food for their families to eat.[18] In actuality, the progressives' plan for slow, natural, and organic food production has been tried. It's called Africa.

Modern technologies allowed our forefathers to leave subsistence living in the countryside to pursue other dreams, and over time, fewer farmers have been feeding more and more people. Less than 2 percent of Americans make a living farming; the remaining 98-plus percent of us enjoy the fruits of a system that generates one of the safest and least expensive food supplies in the history of the world. Americans spend half as much of their disposable income on food today as they did in the 1950s. Not only do we spend less of our income on food, but we eat better, too—even if we don't like to admit it. As John Steinbeck put it when traveling across America in the 1960s, "Even while I protest the assembly-line production of our food, our songs, our language, and eventually our souls, I know that it was a rare home that baked good bread in the old days."[19]

The food police take these changes for granted and appear

oblivious to the challenges of a fast-growing global population that will threaten farmers' ability to feed the world. It is true that many Americans today are more threatened by obesity than starvation, but this is the far lesser of the two evils, especially for the poor among us. And just because we have enough to eat *today* doesn't guarantee the same for the future—especially if we tie the hands of innovators and wait until a crisis is upon us.

Disparaging modern production agriculture, Michael Pollan could conveniently write in 2005 about a corn supply that was "too cheap."[20] Likewise, back in 2002, Marion Nestle argued that "food is too cheap in this country."[21] Little did they know that that only a few years later, in 2008, corn prices would more than double, jumping over 150 percent, even as production rose. As I write this, many commodity prices are near their highest levels in recorded history. In 2011 the *New York Times* published an editorial titled "Food Crisis"—this time not because of the intrusion of evil industrialists into the food chain but because worldwide "food prices are soaring to record levels."[22]

Apparently, the food elite haven't agreed on how to handle their contradictory message. In one corner, we have a cohort sticking with the old message. Writing for the *New York Times* only a few days after the same publication ironically pronounced the crisis of high-priced-food, Mark Bittman asserted that "prices for these foods are unjustifiably low."[23] In the other corner are the food police who tell us, "We are in a food emergency. Speculation and diversion of food to biofuel has contributed to an uncontrolled price rise."[24] So, which is it? Are food prices too high or too low? Are government policy and the agribusiness conspiracy pushing food prices to an unjustifiably high or low level? The good news for the food police is that their recipe calls only for overwrought emotional appeals, not consistency of message.

As recent history illustrates, falling food prices are not a guarantee. We have thankfully avoided the Malthusian nightmare *so far,* but past performance is no guarantee of future success. In *The Great Stagnation,* economist Tyler Cowen demonstrates that the overall rates of productivity growth that we have come to take for granted are diminishing. In agriculture specifically, some commodities, such as wheat, are beginning to show slowdowns in yield growth. Such facts have led some eminent agricultural economists to observe that "on the heels of a recent rapid run-up in global food commodity prices, old doubts have resurfaced about our collective ability to sustain increases in global food supplies that outstrip the growth in demand."[25] I don't put much stock in the doomsday predictions about peak oil and population bombs, but only because of my faith in the ability of humans to innovate and adapt. But if the food police get their way in stifling innovation and impose one-size-fits-all regulation, no doubt about it, prices will rise, and I fear all faith will be lost.

Two decades after the fall of the Soviet Union, we seem to have forgotten the kind of food to be expected from a government empowered to control the food production system. In the 1980s-era Soviet Union, "[s]hortages, food rationing, long lines in stores, and acute poverty were endemic."[26] Because of food shortages, "the average intake of red meat for a Soviet citizen was half of what it had been for a subject of the Czar in 1913." There was no need to ration fruits and vegetables, because there were none. They were "permanently out of stock and thus out of reach for the people of the socialist state."[27] Things were so bad in Russia that Boris Yeltsin's visit to a Houston supermarket in 1990 caused such a "traumatic experience" that it "led him to conclude socialism was a pipe dream and that capitalism might be the only realistic path to prosperity."[28] The food police may not advocate communism or even see themselves as socialists, but their readiness to empower government to control food businesses; to centrally direct agricultural output through heavy taxes, subsidies, and public agency purchasing requirements; and to override consumers' free choice with everything from a gentle nudge to outright ingredient bans is slowly leading us down the road to serfdom. By enacting policies that idle the capitalist engine of our prosperity, the food police are killing the goose laying the golden eggs.

None of this is to say that there aren't problems associated with our modern food system. There are. In many cases, the food police have properly identified adverse consequences of a highly efficient, consumer-driven food production system. The key question is what to do about the problems. Any reasoned approach must be rooted in a belief about the imperfectibility of man and nature; recognize the knowledge and wisdom embodied in emergent institutions and markets; and rejoice in the abundance and prosperity that man has been able to accomplish through his interactions with food—rather than offer complaints about the utopia that we don't have and will never achieve.

The progressive worldview asserts that man in his natural state is basically good. As a result, the world's problems are

not a result of a fallen man but of unjust institutions and social systems. With special knowledge of how the world works,
the progressive elite believe they can reengineer social systems
and usher in a utopia. Once man is freed from unjust socially
constructed institutions, he can flourish. The constraints that
bind us are artificial, institutionally determined results of bad
policy or bigotry.

All this carries over into the food elite's views on nature.
We are told that nature is basically good. As a result, our problems with food are a result not of nature or biology but of unjust institutions and inequitable farming systems. Man should
work with nature, not against it. The lack of food safety, food
security, nutrition, and profitability is a problem not inherent
in nature but with unjust societal systems and institutions.

Yet anyone who has ever raised a two-year-old knows that
man (or, in this case, child) has an ornery streak. Humans,
while capable of great compassion and generosity, will never
be perfect, and Mother Nature can be cruel and unforgiving.
The food abundance we've enjoyed is a triumph of human in-
genuity over nature's indifference to us. There are real trade-
offs that must be made, and scarcity is a reality of our world.
As the author and economist Thomas Sowell reminds us, beware that "many of what are called social problems are differences between the theories of intellectuals and the realities
of the world—differences which many intellectuals interpret to mean that it is the real world that is wrong and needs
changing."[29]

There is perhaps no more critical question than how our
society chooses to relate to food. After all, everyone must eat.
Will we raise a generation of children who look at food with

optimism and innovation, aspiring to be technological innovators such as Louis Pasteur or Norman Borlaug? Or will we raise a cohort of traditionalists who can find nothing better than what is in the past? By cultivating panics and promoting agricultural and environmental idealism, the food police have assumed the moral authority on food. Their message is deceptively appealing because of its traditionalist overtones and fear-mongering about the safety of food. Make no mistake about it, however, nothing short of our ability to feed ourselves and the world is at stake.

Let me be clear: if we are to discuss the morality of food, foremost in our minds must be the ability of people (especially the poor) to eat. Whatever else the modern food elite may *say*—their push for soda and Twinkie taxes, their promotion of local foods, their attempts to regulate processed and fast food, their aversion to biotechnology—their actions make it harder for the poor to meet the most basic necessity of fending off starvation. The irony of many contestants on reality cooking shows such as *Top Chef Masters* is on display for all to see: they pledge to give their winnings to help the poor, all the while promoting a food system that's sucking money out of the back pockets of the least advantaged. Abraham Maslow's famous "hierarchy of needs" had a foundation, and this is it. And yet those of us who oppose the actions of the food police—the same food police who have turned Maslow's triangle on its pointy and now unstable head—are called immoral? Please.

I don't care if you refrain from eating foods made with biotechnology or growth promotants or preservatives. But don't tell me it is immoral to allow farmers to sell consumers these cheaper alternatives. I know the old left-wing canard that it's not fair that poorer consumers can't afford the local, organic veggies bought by the rich. Look—life isn't fair. And until someone can fully articulate an alternative reality where perfectly happy egalitarians, working arm in arm, generate enough wealth to afford the things the rich now can, I regard such arguments as sentimental hogwash. As if people, deep down, *want* equality with the money they've earned. We are, after all, constantly trying to best our neighbors. As David Brooks put it, "We are a democratic, egalitarian people who spend our days desperately trying to climb over each other."[30] Moreover, what arrogance is it to presume that if the poor had $1,000 more in their pocket, their highest priority would be to spend it on heirloom tomatoes? As the *Economist* magazine once put it, "Capitalism can make you well off. And it also leaves you free to be as unhappy as you choose. To ask any more of it would be asking too much."[31]

You don't have to like all modern food technologies, but you should at least have the truth to know what you're giving up—and what the food police will force others to give up—when they're rejected. The hard, cold reality is that we can have expensive, fresh, "natural" food or affordable, processed, modern food. Choose what you will, but don't deny others their choice.

However else it might seem, my critique of the food police is a cry for humility in the face of uncertainty. The world is a complicated place. Anyone who claims truly to know that food was once "too cheap" or is now "too expensive" adorns a crown of arrogance. I have a few hypotheses about why more Americans are overweight than they once were, but I'm not sure we should do anything about it or what might happen if we did. Those who claim otherwise assume a level of competence beyond human limits. There is no food czar (at least not yet, thankfully). There is no top-down planning of our food system. The food elite seek explanations for the ills they see and ascribe malevolence to food companies and corporate farms, engaging in extravagant conspiracy theories. But the food system we have is a result of a complex emergent process that defies the limits of human rationalization. Conservatism, at its best, isn't a knee-jerk reaction against all change, but rather a carefully considered recognition of the wisdom embedded in the unplanned outcomes that have emerged around us as a result of our individual actions. Liberalism, at its best, recognizes that people should have the power to choose for themselves and that we don't have to do things the same way we always have. Between the best of these ideologies lies the right path for food.

The food elite want a supposedly more rational food system that conforms to their vision of the world and to textbook theories of social order. Alas, the real world will not yield to romantic dreams. Look to the outcomes of those who have tried to rationally plan their nations' food production. Stalin and Mao starved millions of Russians and Chinese in an attempt to impose a supposedly more rational economic order. The food system we have is not the result of a single top-down plan, but that doesn't mean it is ill-informed; the food we eat and the prices we pay come about from billions of individual decisions made by consumers and farmers and agribusinesses and grocery stores all trying to make themselves as well off as possible within a framework of norms, rules, and institutions.

Prices aren't tools to be used by a more informed elite to direct the rest of us; they are signals about the world around us. The hubris of the food elite lies in their desire to replace this complex, evolved process embodying an immense amount of information with their own limited vision of the world. The truth is that no one knows enough to tell everyone else how to eat.

Let me tell you what I *do* know. Both my grandmothers raised chickens in their backyards, and wrung a neck when it was suppertime. Unlike today's generation, they didn't farm their backyards to find existential meaning, but out of cruel, dirt-poor necessity. I'm all for people who want to move out to the country and try their hand at hobby farming. More power to them. But don't lecture those who've been there and done that. My grandmother might occasionally reminisce about simpler times, but she remembers well the unplumbed, un-wired house in which she raised my father and his siblings, and those of us that close to a life the food police romanticize and seek to return to, even if by regulatory force, say, "Enough!" We've had our backyard chickens, and Tyson is just fine.

3

FROM COPS TO ROBBERS

A BRIEF HISTORY OF FOOD PROGRESSIVISM

F. A. Hayek lived in Austria in the aftermath of World War I and witnessed the rise of socialism and fascism firsthand. When he later immigrated to England to work at the London School of Economics, he was alarmed to find many of the same intellectual undercurrents at work in Britain that had led to the ravaging of his homeland. Writing in *The Road to Serfdom*, he warned his new countrymen of the fate of such thinking:

> Contemporary events differ from history in that we do not know the results they will produce . . . It would be different if it were given to us to live a second time through the same events with all the knowledge of what we have seen before. How different would things appear to us; how important and often alarming would changes seem that we now scarcely notice! . . . Yet, although

history never quite repeats itself, and just because no development is inevitable, we can in a measure learn from
the past to avoid a repetition of the same process.[1]

Fortunately, the British made a U-turn on the road to
serfdom—perhaps in part because they could see the consequences of their actions in the not-so-distant history of their
neighbors. We, too, might learn a thing or two about where we
are heading by turning back the clock to more fully appreciate the origins of the food police and the unintended consequences of their policies.

Perhaps the best place to start is to see that despite the
fact that exotic organic veggies seem chic, interest in them
has been around for centuries. Indeed, the tension between
traditional grain-growing, mono-crop agriculture and the
lure of alternatives is old news. As the British historian Joan
Thirsk noted in her book *Alternative Agriculture,* animosity
toward conventional grain farming has been around since at
least the fourteenth century.[2] But cries for change throughout history fluctuated with the economy. When growing conventional grains became less profitable, alternatives were
sought. In the cyclical transitions in and out of alternative
agriculture, the food police rode the tides. Each time, the
argument wasn't that more profitable alternatives existed, but
rather, that grain-growing was *wrong* and alternative cropping
systems were good and righteous. Thus, it apparently wasn't
sufficient to change old ways.

Farmers and nobles promoting alternative agriculture
wrapped their decisions in a cloak of morality. Economic incentives are apt to be interpreted as moral imperatives, espe-

cially when we want others to make the same choices we do. Throughout British history, foreign food trade was, at times, seen as both moral and immoral. So, too, was growing grain. And these views correlated almost perfectly with the prevailing economic climate. So, if you find yourself on the receiving end of a condescending sermon by the food police, just wait a few decades and you may find the shoe is on the other foot—if you choose to wear it.

At least the modern-day food police can find some comfort in knowing they have a long heritage. In medieval Britain, a commodity couldn't be exported without a license, and a new crop couldn't be grown without the king's permission. Farmers were required to tithe their commodities to the state church or were otherwise imprisoned. Before modern capitalism, crony capitalists fought to get monopoly rights for growing new crops. And yes, even in the 1700s there were government payments for growing grain. It shouldn't be surprising that some Brits were clamoring to leave this system behind.

When the Pilgrims left England and eventually came to America in the early seventeenth century, they brought with them the ideals of family, individualism, and education that would lay the foundation for the emergence of a new democratic republic a century and a half later. But when it came to food, it seems that puritanical idealism got the better of them.

When the Pilgrims arrived in what came to be called New England, Governor William Bradford set up a communitarian agriculture system. Food and supplies were pooled and then doled out by officials based on equity and need—a system he called common course. And what a beautiful idea it was. The immigrants to the new land were in it together, helping their fellow man, all the while skirting the corrupting influences of the market. Although Karl Marx wouldn't be born for another two centuries, the Pilgrims' agricultural system embodied his ideal, "from each according to his ability, to each according to his need."

No matter how devout were the Pilgrims, they were but mere mortals. Rather than conceding to the planners' good intentions, the layfolk responded to the incentives inherent in the system. When you can't reap the fruits of your labor, why work? Bradford himself wrote in 1647 that the communal system "was found to breed much confusion and discontent," in part because "young men, that were most able and fit for labour and service, did repine that they should spend their time and strength to work for other men's wives and children without any recompense . . . this was thought injustice."

Bradford recognized the problem. Even a tribe so godly and righteous as to pick up, leave home, and traverse an ocean to begin a commune of sorts couldn't be compelled to goodness through cajoling and community spirit. Bradford decided to give each family a private parcel of land from which they reaped all that was sown. The same men and women who were previously chastised for their idleness were now suddenly productive. Bradford wrote that the new system "had very good success, for it made all hands very industrious, so as much more corn was planted than otherwise would have been." The modern food police would have us believe that more neighborliness and community spirit are needed to revolutionize the food system, but while much has changed over the past 350 years, human nature hasn't.[3]

The father of modern economics, Adam Smith, writing

in the late eighteenth century, said it best: "It is not from the benevolence of the butcher, the brewer, or the baker, that we can expect our dinner, but from their regard to their own interest."[4] I certainly wouldn't expect to find something even as mundane as buttered toast for breakfast if I relied on the altruism of the deliveryman, the wheat farmer, the salt miner, the dairyman, and all the other people who must be coordinated to provide the wheat, salt, milk, baking powder, and yeast. Everyone knows this to be true, but the food police unwittingly seek to undermine the profit incentives of the system that make it all happen. Alas, that's unlikely to be the lesson our kids will learn in November when drawing hand-shaped turkeys at school—assuming the food police haven't deemed the craft too degrading to the birds.

The Pilgrims learned another important lesson, one that literally meant the difference between life and death. With help from the Native Americans, they learned how to grow corn. The uncivilized crop grown by uncivilized people did not sit well with settlers accustomed to the wheat and barley of their homeland. Fashion, however, gave way to necessity, and the versatile and productive corn became a staple. Soon the American colonists not only were able to feed themselves but began exporting agricultural commodities back to the Old World.

There is nothing more infuriating to the elite than to witness an event they can't explain—or control. As a result, the cultural elite of the time decried the transition to the unfamiliar corn, and ironically today's cultural elite do the same. Modern food advocates deride the commodity in scathing documentaries such as *King Corn*. Writers such as Michael

Pollan are disgusted with our "river of cheap corn" and are appalled to find remnants of corn in our DNA. With some effort, we might remove corn from our diets, but its influences cannot be whitewashed from history. It is not just our bodies that have been genetically influenced by corn; corn and agriculture have affected the very DNA of our nation itself.

The emerging agricultural basis of the American colonies was not just a way to make a living; it helped define the nation's cultural and political identity. In the years surrounding the Revolutionary War, the colonists had to find an identity separate from the British culture they had previously aspired to emulate, and agriculture provided the means of differentiation. In *A Revolution in Eating,* the historian James McWilliams writes that the rejection of British culture "led Americans to embrace the virtues of the farming life as an explicit cultural and political cause. Thus the virtues of the frontier, as a conscious choice, rather than out of necessity, became an animating force in American life."[5] Americans were simple farmers eating simple food and they were proud of it. As the contemporary Thomas Jefferson put it, "Agriculture . . . is our wisest pursuit, because it will in the end contribute most to real wealth, good morals and happiness." And so began our romance with agriculture.

When the industrial revolution took hold and the process of urbanization began, the changes were thus seen by some as almost un-American. At the time of the Revolution, approximately 90 percent of the American labor force consisted of farmers. By the 1850s the figure had dropped to around 65 percent, and by the early 1900s, farmers made up less than 30 percent of the labor force.[6] With the growth of cities, the rise of textile mills and food manufacturers, the mechanization of farming, and the influx of immigrants, it seemed that a way of life that had defined a nation was being lost. The technical changes that were afoot set the stage for the rise of progressivism and the first epic of the American food police.

The Progressive movement that began in the late 1800s was intimately interested in food and agriculture. In fact, the long-standing importance that progressives have placed on food and health is often underappreciated—as is their paternalistic attempts to influence what others eat.

The Progressive movement can be seen, in part, as an outgrowth of Christian social gospel. It is perhaps no mistake, then, that in the late 1800s the two Kellogg brothers, following practices of the Seventh-Day Adventist Church, established what would eventually become known as the Battle Creek Sanitarium. There, such famous patients as Henry Ford and President Warren G. Harding obtained spiritual and physical healing through dietary reform and a host of therapeutic treatments, including frequent enemas. The Kelloggs encouraged exercise and hygiene, but became best known for their promotion of a low-fat, low-protein, high-fiber diet, which was embodied in their new flaked breakfast cereal. The Kelloggs were just a small part of the larger grassroots pure foods movement.

While the Kelloggs were trying to convert all who would listen to their preaching on the latest dietary fad, others were intent on reforming the whole country—by force, if necessary. A confluence of economic and political events in the 1920s gave that period's food police the leverage to outlaw the sale of alcohol.[7] Prohibition became entangled with the suffrage movement, the establishment of the income tax, religious fervor, and anti-immigration sentiment. Class paternalism was a key motivator for Prohibition, and Progressives were at the forefront of the movement to save lower-class drinkers from their own defective judgment. Some leftists were motivated by the belief that liquor was a tool used by the capitalists to keep the working man in his place. If the capitalists were at fault, I'm not sure who was to blame when people took to the speakeasies and backwoods stills of their own volition.

Rather than ushering in a sober utopia, Prohibition yielded all sorts of chaos and unintended consequences. Banning alcohol only increased the incentive for the likes of Al Capone to provide what customers wanted, only now underground and in a deadlier fashion. The economic foundations of towns such as St. Louis and Milwaukee were devastated. Those with political connections enriched themselves and went their merry way. Religious groups were still allowed to use wine for sacramental purposes, and one rabbi in Los Angeles found a new devotion to the faith—with his congregation increasing from twenty to a thousand families after Prohibition went into effect. There were more than a few O'Malleys and O'Connells claiming Jewish heritage. Policemen willing to turn a blind eye suddenly found their bank accounts ballooning.[8] Fortunately for those who wanted to blow off a little steam, state and federal governments realized they needed the liquor taxes, and Roosevelt repealed Prohibition, concluding, "I think this would be a good time for a beer."

I can see it now: some future president thirty years from now—after repealing the fat, soda, and meat taxes, the salt and saturated fat bans, and the prohibition against food traveling more than a hundred miles—saying, "I think this would be a good time for a Happy Meal." That is, if there is still someone around able to sell her one.

While Prohibitionists were concerned with the meal's beverage, others took aim at the main course. Many high school English students will remember Upton Sinclair's 1906 book *The Jungle*, about the Chicago meatpacking industry. Fewer will probably remember that Sinclair wasn't primarily concerned about the meat. *The Jungle* overtly promoted socialism, with the book's main character eventually converting to the cause and going by the name Comrade Jurgis. Published eight years before the First World War and a decade before the Bolshevik Revolution, the book featured characters who could confidently promise a socialist utopia, claiming among other things an "unlimited food supply" and that "after the triumph of the international proletariat, war would of course be inconceivable."[9] It also bears mention that Sinclair was an ardent supporter of Prohibition.

Despite a failed run for the governorship of California, Sinclair could still take some pride in knowing he'd caused enough public outcry to help get new food safety regulations through Congress. Alas, he was dissatisfied that he managed "only" to clean up the meatpacking plants. A dismayed Sinclair said, "I aimed at the public's heart, and by accident I hit it in the stomach."[10] His stomach punch helped lead to the first major federal food regulations and the creation of the precursor to the modern-day FDA. Ironically, this is the same FDA that modern-day progressives deride and the same meatpacking industry criticized in present-day books such as *Fast Food Nation*. At least Sinclair was upfront with his politics.

Although Sinclair complained about collusive meatpackers,

it wasn't until 1920 that Progressives were able to dissolve the National Packing Company. Federal efforts to prevent monopoly power had been around since the Progressive Woodrow Wilson sought to curb food inflation during World War I. During that time, he had the Federal Trade Commission investigate the agricultural industry from "hoof to the table."

What exactly has resulted from the century-long attack on the meatpacking industry? Despite decades of investigation and regulation, the meatpacking industry today remains dominated by a few large firms, just as it was in Upton Sinclair's time. Four companies are today responsible for 80 percent of the beef processed in the United States.[11] And the industry probably employs as many immigrants as it did in Sinclair's time. Killing is nasty business no matter who does it, and it was only slightly less pleasant in 1900 than it is in 2012. As long as there are those who wish to eat meat, there will be those willing to pay others to kill. And until we find some better way to do it, employment in a slaughterhouse provides the best alternative for some people working their way up the economic ladder.

The fact that most of the meat processed in this country comes from a few large firms says more about the way the costs of processing beef change with the scale of production than it does about all the Progressives' attempts to hinder cost-cutting efforts. The research shows that any market power these firms have gained through consolidation has been far more than offset by technological innovations and cost-cutting measures.[12] We have to remember, after all, that these "evil" meatpackers have somehow found a way of selling us beef and pork that is now 20 percent less expensive and poultry that is 50 percent less expensive than it was in 1970.[13] And it is safer, leaner, and of more consistent quality, too.

As evidenced by the creation of food safety standards, not all changes in the Progressive era were counterproductive, and there was at least one other development that had a positive impact on modern-day agriculture. Justin Morrill, a congressman from Vermont, unable to attend college because his father could not afford the tuition, became an outspoken advocate for low-cost college education. In the late 1800s, Morrill helped enact the Land Grant Act, which provided for the establishment of publicly funded agricultural and technical educational institutions across the nation. The act was followed by legislation providing federal funds for agricultural research and education. These developments eventually led to the creation of universities and agricultural experiment stations from Cornell University to the University of Minnesota to the University of California–Berkeley. These Land Grant universities created new, higher-yielding plant varieties and more productive and sustainable cropping systems, and spread the new scientific findings about food and agriculture all over the countryside. The benefits of these investments have been substantial, with current estimates indicating that the benefits of the research have exceeded their costs by a factor of 32 to 1.[14] Economists estimate the rate of return on agricultural research at between 10 and 50 percent.[15] Try consistently getting that in the stock market! And no matter what you've been told, the primary beneficiaries of this research haven't been farmers, industrialists, or agribusinesses—but you and me, the food consumers.[16] Yet the food police's most ardent wish is to strip us of these benefits and have us pay more for food.

If America seemed to be on a path toward a more natural and less industrialized food production system, the Great Depression and the Second World War would change all that as the country recognized the need for a production system that could feed the nation in a cost-effective manner. Spawned by the innovations coming from entrepreneurs and the agricultural colleges, dramatic productivity increases began to occur. In the early 1900s a farmer could expect to harvest about 29 bushels per acre of corn planted. With the development of hybrid seeds and the application of synthetic fertilizer, average corn yields increased to about 40 bushels per acre in the 1950s, and jumped to 113 bushels per acre in the 1980s. Today, with the use of biotechnology, average corn yields are around 130 bushels per acre, and yields of close to 300 bushels per acre can even be attained in ideal conditions.

It wasn't just corn; the same thing was happening in all facets of agriculture. In the 1960s a sow could be expected to produce about 1,400 pounds of pork through the offspring she birthed. Today the average sow will produce about 3,600 pounds of pork. The result is that food is now *much* cheaper than it once was. As a percentage of their income, Americans today spend half as much as they did on food a hundred years ago. Of course, incomes have risen dramatically over that time as well—meaning that the real price of food has substantially fallen. It took a schoolteacher in 1900 a little more than one hour's worth of work to earn enough to buy a dozen eggs. Today's schoolteacher has to work less than three minutes to earn enough to buy a dozen eggs.[17]

Dramatically higher levels of productivity imply that today's farmers (and, by implication, today's consumers) get much more food using many fewer inputs. That's the signal that falling food prices send us: we're using up less of the things we all deem valuable. Despite the fact that agricultural output is 2.75 times higher than in 1950, U.S. Department of Agriculture data reveal that agricultural input use has remained essentially unchanged.[18] That's like a technological innovation enabling me to write almost three books in the same time it now takes me to write one. These changes imply that farmers can now produce a lot more food with a lot less land. Indeed, despite now having a lot more to eat, agricultural land use is down 27.5 percent since 1950. One would expect this development to be a source of great pleasure and pride for the modern food progressive worried about the environment and the poor, but the food police are anything but happy about the way we eat.

The U.S. farm population peaked around 1910, and after the First World War, commodity prices fell precipitously as European agricultural production resumed. Progressives of the era lacked the power to enact the price controls they desired, but the intellectual framework provided by the Progressive era enabled the New Deal policies of the 1930s. It wasn't until the Great Depression that Franklin Roosevelt enacted farm price supports and supply controls in an effort to better the lot of farmers.

The Progressives' efforts eventually led to outright destruction of agricultural commodities at a time when much of the nation was starving. One of the most notable cases was that of Ohio farmer Roscoe Filburn, who was taken on by the U.S. government for the heinous act of growing too much wheat. Even though he planned to use the wheat only on his own farm, the Supreme Court decided that his actions had violated the Agricultural Adjustment Act of 1938. His sin? By growing more wheat than his allotment, the court claimed, he had indirectly pushed the price below the bureaucratically determined minimum. The solution was simple: his wheat had to be destroyed.[19] This kind of inanity wasn't limited to court cases. The earlier Agricultural Adjustment Act of 1933 paid farmers to destroy hogs, and millions of dollars were paid out to cotton farmers to plow their crop back into the ground!

So, despite the lines at the soup kitchens during the Depression, the planners of the era deemed it morally acceptable to destroy food in an attempt to ratchet up farm prices. Fortunately the government no longer destroys commodities to control supply, but many of the price and income supports that arose as part of the New Deal remain, in convoluted forms, today. Ironically, today's food progressives abhor farm programs and blame them for everything from obesity and diabetes to the massive growth in farm size and rural outmigration.

One would think the take-away lesson for progressives is that the results of government planning and control rarely match the good that was intended. Yet progressives apparently believe that the problem is one not of too much government control but of too little. In his pitch to keep Republicans from regaining office, Obama made this analogy for the recent economic downturn: Why would you give the keys back to a guy who has just driven his car into a ditch? The first generation of food police banned alcohol, destroyed crops, dismantled agribusinesses, sued farmers for making their own planting decisions, enacted obstinate farm subsidies, and issued wartime food ration booklets. Do we really want to hand them back the keys?

HIPPIE FOOD AND THE HEIRS OF THE PROGRESSIVE FOOD MOVEMENT

Like the first Progressives, the 1960s generation was poised for a revolt against food industrialization. Those seeking peace and love also needed a new diet to fit their worldview. Generations removed from production agriculture, the hippies rebelled against the prevailing system and went back to the farm to grow and eat food that wasn't being manipulated by the Man. Communes, food co-ops, and organics came to symbolize hippies' dropping out of society and expressing their distrust in the prevailing economic order.

This was a generation raised in the atomic age. They were primed for fear, and that's what they got. Rachel Carson's *Silent Spring* made them fearful of pesticides and environmental degradation, Ruth Harrison's *Animal Machines* sparked paranoia about the treatment of farm animals, and Paul Ehrlich's *The Population Bomb* raised alarm about overpopulation and the mass starvation of humans.

The movement rejected technology as a potential solution to the problems; after all, it was technology that had created the nuclear weapons underpinning the fears they faced. Moreover, the hippies were rejecting a capitalist system that benefited the creators and owners of new technologies. At least with respect to food and agriculture, the hippies were becoming Luddites, and they were anticapitalist by creed.

They rediscovered organic food, which has its roots in the natural and organic food movements in Germany. In the 1920s, after German scientists discovered a process to create synthetic nitrogen and thereby drastically improve crop yields at a much lower cost, a German philosopher and mystic, Rudolf Steiner, became appalled at this "unnatural" development and reacted by encouraging "biodynamic," or organic, food production. The mystic and "natural" aura surrounding the process later engendered the hippies' embrace of organic food.[20]

Although the hippies were a subculture, they had significant influence on American culture writ large. Not only did they influence a generation of teenagers to wear bell bottoms and listen to the Grateful Dead, but they affected the way Americans thought about food. A previous generation that had operated under the implicit assumption that newer food technologies were better—TV dinners, Tang, canned biscuits—now began questioning that wisdom. Although tie-dyed T-shirts soon went out of fashion, it took time for the fashion of "natural" and organic foods to take hold. Now organic has gone mainstream.

In fact, organic has gone too mainstream for the elite wing of the modern food progressives. Whatever we were told about the benefits of organic, it seems that it wasn't actually about the food after all. In his bestselling book *The Omnivore's Dilemma*, Michael Pollan subtly makes this case by "exposing" the size of the modern organic farms. Of course, the problem with organic labeling standards, now defined by the USDA and FDA, is that they have nothing to do with social justice, small farms, or local production. The hippies promoted organic only to find now that they don't like what they created. Just like all religions entering a modern world, organic-ism is undergoing its own reformation. Disciples are now converting to the locavore-ist, vegetarian, terra-ist, or slow-food-ist denominations.

The modern food progressives are no longer disgruntled teenagers. They are now influential journalists, academics, regulators, and politicians. Their advocacy has led to a garden on the White House lawn, a new mandate for the FDA, Michelle Obama's Childhood Nutrition Act, and a host of new proposals to regulate food and health. Unlike with the Progressives of the early 1900s, however, the scientific and technological optimism is gone. Today's progressives want a revolt in food, a retrograde revolution seeking to return food and agriculture to some romanticized past. And to accomplish their objective, they have spread anxiety and fear about our modern food system and created a caricature that cannot withstand scrutiny.

How did we reach such a state of affairs? The American food system was once considered the envy of the world—we produced more, innovated more, and ate better and more cheaply than the citizens of almost every other country in the world. And by and large, we still do, no matter what romanticism is expressed by progressives about European food. It is common to hear American travelers in Europe remark, "We don't have anything like this in the States," but as the European politician Daniel Hannan retorted, "Yes you do: They're called restaurants, for heaven's sake."[21]

Before we accept the premise that our food system is in need of revolution, it is prudent to look back, see where we've been, and recognize how far we've come. In the midst of the industrial revolution, self-described socialists wished to put an end to capitalist-driven industrialization by overthrowing the bourgeoisie. But they, like the food police today, failed to appreciate the incredibly positive change brought about by technological development and the capitalistic system. In her book *Grand Pursuit*, Sylvia Nasar beautifully articulates what escaped their notice:

> For a dozen centuries, as empires rose and fell and the wealth of nations waxed and waned, the earth's thin and scattered population had grown by tiny increments. What remained essentially unchanged were man's material circumstances, circumstances that guaranteed that life would remain miserable for the vast majority. Within two or three generations, the industrial revolution demonstrated that the wealth of a nation could grow by multiples rather than percentages. It had challenged the most basic premise of human existence: man's subservience to nature and its harsh dictates. Prometheus stole fire from the gods, but the industrial revolution encouraged man to seize the controls.[22]

Over the past two centuries, we have witnessed what can only be described as a miraculous transformation in the safety, quality, and quantity of the food we eat. Yet, at least in the realm of agriculture, it seems the food police would tie our hands, wrest the controls from the engine of growth, give the gods back their fire, and let nature run its course. With history as our guide, let us not yield to the demands of the food police but anxiously await what more the technological innovators have to offer.

4

ARE YOU SMART ENOUGH TO
KNOW WHAT TO EAT?

We make about two hundred food-related decisions each day.[1] With so much practice, isn't it obvious that we are capable of knowing what to eat? Not so, according to the food police.

When it comes to the healthfulness of our dietary choices, we are apparently unable to think clearly. Our wants and desires must be transformed. When answering a question about what the government is doing to reduce food waste and make nutritious foods more appealing, Tom Vilsack, Obama's secretary of agriculture, told a crowd, "It's going to take time for people's taste to adjust . . . So, we have to make sure that what we do is create the appropriate transition."[2] I wasn't aware that my taste buds needed a transition.

In a recent speech promoting her Let's Move! anti-childhood-obesity campaign, Michelle Obama said, "It's not about telling people what to do."[3] That's normally what people say right before they try to guilt you into doing what they want you to do. It does make one wonder what the First Lady's campaign is all about if she isn't trying to tell us to do *something*.

It is not just that the food police want us to make different food choices—advertisers, after all, routinely try to accomplish this feat—it's that our choices are deemed incorrect or morally defunct and must be changed even if it means using the government's regulatory power to do so. Whereas we can choose to ignore the advice of preachers or advertisers, the same cannot be said of regulators.

To see how the food police are gaining the helm of power, we have to turn to the halls of academia and to the seemingly obscure work of a group of psychologists seeking to rewrite the rules of economics with a field of study called behavioral economics.[4] Why worry about some academic squabbling? Because behavioral economics has provided the philosophical basis and the real-world traction for supplanting our own preferences and beliefs, as revealed in our individual choices, with those of the food elite. If you want to know why food paternalism will be around a long time after the Obama administration is gone, you have to understand that the elite have created a pseudoscientific theory for why their helping hand is needed.

The influence of behavioral economics should not be underestimated. It has reached such influence in the food policymaking arena that the USDA held a conference in 2010 entitled Incorporating Behavioral Economics into Federal Food and Nutrition Policy and has published reports with titles such as "Could Behavioral Economics Help Improve Diet Quality for Nutrition Assistance Program Participants?"[5]

Even economists who tend to be sympathetic to consumer sovereignty have begun to argue with points such as "behavioral economics . . . may expand the scope of paternalistic regulation."[6] With the popularity of bestselling books such as *Predictably Irrational* and *Nudge,* behavioral economics has gone mainstream. The ideas have permeated the highest levels of regulatory decision making at the USDA and the FDA, and one of the authors of *Nudge* was recently appointed as President Obama's regulatory czar. Make no mistake about it, behavioral economics is the engine behind the new food paternalism.

My critique of the specific food policies proffered by the food police will come later in the book. In this chapter, I take on a powerful emerging motivation underlying the call for new food policies. It is the idea that you are unable to make wise choices for yourself. If you cannot choose wisely, then the food police have all the motivation they need to decide for you. The paternalistic food police will cite scientific studies to tell you why we need Uncle Sam's helping hand. We'll see why those same studies imply no such thing.

WHAT IS BEHAVIORAL ECONOMICS?

Behavioral economics shows that people are . . . well, human. Behavioral economics suggests that we are irrational and biased; that, on some level, we are incapable of making coherent choices that serve our best interest. If we cannot act in our own best interest, then the food police believe they can step in and make our lives better.

So, what is the evidence that people cannot choose wisely for themselves? Some of the most interesting examples can be found in Brian Wansink's fascinating book *Mindless Eating*. Wansink shows that our eating decisions are influenced by subtle cues that seemingly defy logic. We eat more jelly beans the more colorful the bowl; we eat more food and stay at the restaurant longer when we think we are drinking wine from California rather than North Dakota, even though both wines are the same; and we often consume more calories when eating from packages with low-fat labels. If asked the question "When do you decide to stop eating?," most of us will give answers such as "When I'm full." Wansink's research shows that we're full of it.

In one ingenious experiment, Wansink asked a group of college students to join him for a free lunch. Half the participants were seated in front of a normal soup bowl, and the other half were seated in front of a soup bowl that (unbeknownst to them) was being continually refilled by a tube connected to a large vat underneath the table.

If we stop eating when we're full, one would expect both groups of students to have eaten roughly the same amount of soup. But Wansink found that those eating out of the bottomless bowl ate 73 percent more soup than those eating from the regular bowl. Surprisingly, the students eating from the bottomless soup bowl thought they had eaten about the same number of calories (and rated themselves as being just about as full) as the students eating from the regular bowl—despite the fact that they had eaten 113 more calories on average! Apparently we stop eating when prompted by some external cues (such as when our bowl is empty or when our plate is clean) rather than entirely internal cues (such as when we feel full).

One of the biggest problems we face when deciding what

to eat is the tough trade-off between choosing to eat something tasty (but unhealthy) *today* with the increased risk of a
heart attack or weight gain at some point in the *future*. Current benefits must be weighed against future potential costs.
The research in behavioral economics shows that we tend to
be overly impatient. We make decisions that seem to suggest
we will always plan to start a diet *tomorrow*. When tomorrow
becomes today, we find ourselves in the same position, choosing to forgo the diet yet another day, telling ourselves all along
that tomorrow will be the day the diet begins. It is as though
our current self and future self can't quite get along.

These kinds of findings have led behavioral economists
to call for a host of paternalistic regulations to override our
preferences. Nobel Prize–winner George Akerlof doesn't even
want to give us a choice in the matter. He says, "Individuals
may be made better off if their options are limited and their
choices constrained."[7] Writing in *Nudge*, Thaler and Sunstein
say they want to give government the power to nudge us into
making the decisions some bureaucrat deems desirable. They
seemingly want the government to assume the power to tell
restaurants how their menus should look and tell cafeterias
how they should be organized. They justify their paternalism
by saying that it is "quite fantastic to suggest that everyone
is choosing the right diet, or a diet that is preferable to what
might be produced with a few nudges."[8] A USDA report says
behavioral economics "expands the array of possible ideas"
for new food policies and can license the government to alter
"elements of the product, such as package size and shape,
the amount of variety, the number of calories, or the default
options on a menu."[9] Writing for *Forbes*, Duke University

professor Dan Ariely says, "Behavioral economics could help us take steps toward designing a better world."[10] I don't know about you, but I'm more frightened than comforted by a psychologist confidently asserting he can *design* a better world with new regulations.

It would be absurd to assert that we never make foolish decisions that, in retrospect, we regret. But despite what we've been told by those Ivy Leaguers who have traversed the well-worn highway between academia and Washington, behavioral economics cannot justify the actions of the food police.

THREE MYTHS OF BEHAVIORAL ECONOMICS

Myth 1: The findings from behavioral economics are really important.

If you want a depressing afternoon, spend some time scouring the psychological and behavioral economics literature. There you will find that we are prone to a plethora of decision-making biases that are apparently limited only by the researchers' abilities to find catchy new names for them. You might begin to wonder how it is we are even able to make it out of bed in the morning and find our way to work. If we are *truly* as irrational as the behavioralists suggest, wouldn't they have to have some way of accounting for the success and prosperity most Americans enjoy? It is a bit odd that we are just now getting around to finding out that our decision-making abilities are so poor that new regulations are needed to improve our lot in life. If we aren't truly more innately irrational than our grandparents, how can the case be made that there is a *new* need for paternalism?

The answer lies, in large part, in the peculiarities of the academic publishing world. We academics are rewarded for novel ideas. One cannot publish by repeating what everyone already knows. The rise in behavioral economics can be explained in large part because the field offered an alternative: a novelty. But it is important to keep in mind that what sells in academic journals does not necessarily possess real-world importance.

Consider a striking comparison. At about the same time behavioral economists were making headway arguing that people were *ir*rational, a different group of economists were arguing that animals (rats, pigeons, and pigs) were rational. Whereas leading behavioral economists wrote papers concluding that people can't figure out how to make themselves better off, another group of economists publishing in top-tier economics journals concluded that, of all things, animals can![11] The counterintuitive claims that were novel enough to fly in academic journals were that animals were rational and people were not. But no one in his right mind would claim that rats are more rational than humans, no matter what was published in peer-reviewed journals.

If the findings of behavioral economics are really as powerful and influential as claimed, then an entrepreneur should be able to profit from the mistakes consumers make. It might not be surprising, then, to learn that Richard Thaler, one of the pioneers of behavioral economics, is a partner in an asset management company that makes investments based on the premise that "[i]nvestors make mental mistakes that can cause stocks to be mispriced."[12] Note that this is the same Thaler who coauthored *Nudge,* which argues for regulators to use knowledge of decision-making biases to improve society's welfare. It is a little unclear whether Thaler believes that our decision-making problems represent an opportunity for him to extract money from us or improve our lives, but at least we know he's confident we often don't choose well.

Alas, a Morningstar analyst concluded, "This mutual fund's distinctive approach may not be enough," in reference to one of the behavioral funds Thaler advised.[13] More generally, research into dozens of mutual funds constructed to capitalize on traders' supposed decision-making problems has concluded that "there is no conclusive evidence to suggest that these strategies outperform [a respective benchmark]"[14] and that "behavioral [economics] mutual funds are relatively synonymous with value investing and not much more."[15] Believers in behavioral economics can put their money where their mouth is by investing in funds designed to capitalize on investors' decision-making biases. Here's my advice: don't do it. And if traders with hundreds of millions of dollars on the line can't put the behavioral economics research to work to earn a buck, what makes you think bureaucrats and politicians can?

The failure of behavioral economics to generate superior performance in consequential matters, such as making money in the stock market, has not prevented some from taking simple findings from academic experiments and drawing grandiose claims about the promise of behavioral economics for creating new regulations to improve our health.

For example, in his bestselling book *Predictably Irrational,* Dan Ariely describes some creative studies he conducted with college students and Halloween trick-or-treaters showing that choices were highly influenced (seemingly irrationally so) by whether a chocolate was advertised as free. Based on the be-havior of the preteen ghosts and goblins, Ariely concluded, "We can use FREE! to drive social policy." He suggests, "Want people to do the right thing—in terms of getting regular colo-noscopies, mammograms, cholesterol checks, diabetes checks and such? Don't just decrease the cost (by decreasing the co-pay). Make these critical procedures FREE!"[16] Notice the statement "Want people to do the right thing"? The implicit premise is that we are either unwilling or unable to make the "right" food and health care decisions on our own. But even if we could get everyone to check his cholesterol, wouldn't we need to know how much it costs before we could say the deci-sion was "right"?

In another set of experiments, Ariely shows how college students reacted to different opportunities to set class proj-ect deadlines. Then, without showing that regular people will respond to health care interventions in the same way that stu-dents responded to assignment deadlines, he makes the leap to conclude, "Think how many serious health problems could be caught if they were diagnosed early. Think how much cost could be cut from health-care spending . . . So how do we fix the problem? Well, we could have a dictatorial solution, in which the state (in the Orwellian sense) would dictate our reg-ular checkups. That approach worked well with my students." While Ariely explicitly recognizes the heavy-handedness of his proposal, he goes on to defend it by encouraging us to "think of the other dictates that society imposes on us for our own good."[17]

Ariely is a first-rate scholar and perhaps we should cut him some slack for trying to publicize his research, but when the elite turn around and use the same book as "scientific evidence" to support new paternalistic food and health regulations, surely we must take pause.

Even some leading behavioral economists agree that the government will be unable to use the findings of behavioral economics to substantively impact the nation's health. George Loewenstein, a prominent behavioral economist and professor at Carnegie Mellon University, surprised many people when he wrote an op-ed for the *New York Times* admitting that behavioral economics cannot reverse the rise in obesity. He said, "For all of its insights, behavioral economics alone is not a viable alternative to the kinds of far-reaching policies we need to tackle our nation's challenges."[18] It seems Loewenstein believes some type of regulation is needed to "correct" our food choices, but the findings of behavioral economics lack the oomph to nudge people in the right direction.

Myth 2: Paternalism is different from elitism.

I have a confession to make. I'm a paternalist. With two elementary school-aged children at home, I don't have much choice, unless I'm willing to let the Wii run 24-7 and allow my kids' teeth to rot from Pop-Tarts and Twizzlers. Yep, I think I know more about what will affect my kids' future happiness than they do, and I exercise my parental authority to make it happen. Although I'm a paternalist with my kids, few would say that makes me an elitist. Why? Because, as Bill Cosby used to say about his children, "I brought you into this world and I can take you out of it."

Whether by law or by social custom, I have a nearly universally agreed-upon right to be master of my children (at least within some reasonably wide legal boundaries). But since when did Richard Thaler or Cass Sunstein or Dan Ariely or any number of the other folks who apparently think they know what is better for me gain the right to make choices on my behalf? When did I become someone else's responsibility? Assuming the right to care for someone when they haven't asked for it is only slightly shy of tyranny.

When paternalistic policies are advocated on regular people such as you and me, it becomes clear that the food elite seek to impose their particular notion of morality on others. By trying to supplant each of our own individual decisions, made in different contexts with our own insights into the options facing us, with a uniform plan of what they believe we should choose, the food police reveal the height of their elitism.

For example, many of the food police want to follow the example of the European Union and ban ranchers and farmers from using growth hormones in dairy and beef cattle production. "Good for them," you might say. After all, aren't these added hormones the cause of the declining age of puberty among American girls? Hardly.[19] For those folks who want to avoid added hormones in meat and milk, they are free to pay higher food prices for the organic varieties.[20] The rest of us can take comfort in knowing that fears over the cost-saving technology are vastly overblown.

A one-pound burger from a steer administered growth hormones has about 3 nanograms more estrogen (the primary hormone compound used to promote growth) than cattle who have never received the added hormones (a nanogram is one billionth of a gram). Yet the same amount of raw cabbage contains 10,896 nanograms of estrogen, and the same amount of soybean oil contains almost 1 million nanograms of estrogen! A birth control pill has many thousands of times more estrogen than that contained in a serving of beef.[21] You might still want to eschew burgers from cattle fed growth hormones, but clearly it would help to have a bit of perspective.

Rather than letting each of us weigh the evidence and decide if the extra cost of organic beef is worth the minuscule increase in risk, the food police want to make the choice for us. Here is the irony. The behavioral economists have told us for years that humans make mistakes by exaggerating the importance of low-probability risks.[22] Yet I have not seen a single behavioral economist use this insight to tell the food police to relax and put their fears about growth hormones, genetically modified food, or pesticides into perspective. Instead, we see the behavioral economists partner with the food police to advocate policies *they* want even if it means ignoring the implications of their own research. The truth is that the many millions of decisions we shoppers make at the grocery store every day reveal that we've decided it simply isn't worth exploding the grocery bill to have beef from cattle that haven't been given added growth hormones. The food police have determined that all these decisions are wrong, and they have no justification for the conclusion except their own desire to remake the world as they want it.

If we no longer use an individual's choices to define what is or is not "good," then whose do we use? Many behavioral economists attempt to skirt this subtle but critically important point. In *Nudge,* Thaler and Sunstein call for paternalistic policies "to influence choices in a way that makes choosers better off, as judged by themselves." Is this not a logical inconsistency? If people's preferences are incoherent, as Thaler and Sunstein argue they often are, then people cannot "judge for themselves." Thus, the elite seek to replace each individual's judgment of the "good" with their own. Paternalists aim to remake the world *as they want it*—a world that differs from the one people have actually created through their own individual actions. According to Kerry Smith, a leading environmental economist and member of the National Academy of Sciences, this is "simply an effort by one group to convince others they are really the 'smart folks' and should make decisions for others in the truly important, complex situations."[23]

Recall that many behavioral economists contend that we do not have sufficient self-control to take care of our future selves. The supposed proof of this irrational behavior is said to be found in survey responses in which we say we wished we weighed less or saved more. But our current self will *always* wish that our previous self had dieted and saved more, because we are now in the position to reap the benefits without paying any of the costs. The paternalist has simply decided that your abstract future self is right and your current-acting self is wrong, and the only possible excuse the paternalist can give for his paternalism is his own preference for your actions.

Paternalists routinely point to the bad outcomes that accrue from unhealthy diets to justify their regulations. Seeing someone with diabetes or heart disease is supposedly proof of poor choices. Such conclusions confuse the undesirability of a bad outcome with a potentially reasonable choice made amid uncertainty. It is not as though anyone actually wants to have a heart attack. But we don't choose whether we have a heart attack; we make choices that influence the *likelihood* of having a heart attack.

In real life, we face choices such as whether we want a piece of chocolate cake that will certainly be tasty now but will increase the chances of a heart attack by, say, 0.00001 percent five years from now. It would make no more sense for the paternalist to call the choice of cake wrong than it would be for him to say it is wrong for you to play a lottery in which you have a 99.9 percent chance of winning $10,000 and a 0.01 percent chance of losing $10,000—even if you unluckily lose the money.

By saying that a heart attack is evidence of poor decision making, the paternalist is drawing a false comparison between the choice made by a fictitious, all-knowing being with perfect foresight and a real human faced with trade-offs and uncertain outcomes. By claiming to know how our dietary choices should be made, the elite are assuming the role of the all-knowing being with perfect foresight. As the economist Ludwig von Mises put it in 1949, "No man is qualified to declare what would make another man happier or less discontented. The critic either tells us what he believes he would aim at if he were in the place of his fellow; or, in dictatorial arrogance blithely disposing of this fellow's will and aspirations, declares what conditions of this other man would better suit himself, the critic."[24]

Given the almost obvious elitism, how is it that behavioral economists have been so effective selling books and promoting paternalistic policies? In part, their success can be explained by the use of a rhetorical illusion. Popular writings on behavioral economics place the reader in the role of the elite, of the person who must decide what to do about all the seemingly irrational behavior. In the book *Nudge,* we are asked to put ourselves in the role of a *choice architect.* Of course, the reality is that very few of us will ever actually be in the position to invoke the paternalistic plans we might scheme. An honest appraisal of paternalism would put the shoe on the other foot. Paternalism doesn't seem so bad when we envision ourselves as the ones with the power to pass laws overriding others' decisions with our own preferences. The appeal quickly fades when we realize that it is most likely our own preferences that will be overridden by someone else. I also suspect the more manly sounding paternalism would lose much of its appeal if it went by its more accurate motherly sounding moniker: maternalism.

Myth 3: Experts can make better decisions than layfolk.

The food police tell us that we need experts and elites to enact regulations to help people avoid bad choices. But aren't experts people, too?

The elites, of course, answer no. They aren't *regular* people. The very concept of being elite implies being more enlightened—above the thinking and mistakes of the regular Joe. They claim to have special knowledge and insight that give them the responsibility and the authority to enact their preferred plans for others' lives. It is true that many of the elite have PhDs from Ivy League institutions. But the knowledge of the elite tends to be a special kind—an academic knowledge about abstract concepts and theories. This *special knowledge* can be contrasted with what Thomas Sowell calls *mundane knowledge*, the more practical knowledge of how things work in the real world; the knowledge of "plumbing, carpentry, or baseball." The elites' special knowledge is only one kind of knowledge, and one that is particularly impervious to refutation by real-world facts. When a plumber's mundane knowledge is wrong, the feedback is immediate (and wet!), but a sociologist or psychologist may never receive any consequential feedback on the veracity of his conjectures.

Sowell argued that "the fatal misstep of such intellectuals is assuming that superior ability within a particular realm can be generalized as superior wisdom or morality over all." Even though special knowledge is often held in higher esteem in our society than mundane knowledge, "in the aggregate, mundane knowledge can vastly outweigh the special knowledge of elites, both in its amount and in its consequences."[25]

Elites mistakenly believe they know more than they actually do, and successes within their own specialties of knowledge lead them to overestimate their competence in other areas—such as policy setting. Are we really to believe that a nutritionist who measures vitamin content in carrots has any keen insight into what will happen within and beyond agriculture when the government subsidizes conversion to organics? The problem is that those with expertise in Shakespeare or string theory fail to realize their lack of knowledge in other areas. With regard to food, those other areas are just as important in determining how people will respond to government action.

This arrogance leads elites to overlook their own decision-making problems in areas outside their focus. Chess players, for example, while able to solve exquisitely complicated reasoning problems on the chess board, utterly fail at solving similar reasoning problems outside of chess. The same is true of world-class poker players and professional soccer stars.[26] Expertise is typically limited to a very narrow field of inquiry. This is why we should bristle when we read of experiments about free candy bars being used to justify health care interventions. At least when a chess player errs, his mistake only costs him the game. When the academic elite lobbying for more paternalistic food polices err, we all suffer.

The hard, cold reality is that experts make mistakes, too. One of the central themes in Thaler and Sunstein's book, *Nudge,* is that we tend to suffer from a *status quo bias.* We tend to pick the option that we know, the status quo, for no reason other than that it is the default. How ironic it is, then, that government itself is perhaps more prone to the status quo bias than are individuals. There is a well-known adage in Washington that it is much easier to pass a new law than to remove an existing subsidy. Today even most environmentalists agree that ethanol subsidies are a bad idea, and yet they continue to exist for no better reason than that it is the status quo.

One of the big problems with experts is overconfidence. After analyzing decades of expert forecasts, psychologist Philip Tetlock concludes that, by and large, experts' predictions are no more accurate than the guesses of a chimp. As Tetlock reveals, the experts aren't discouraged by their fallibilities; they tend to be *over*confident and in many cases dogmatic.[27] So much for relying on the experts and their supposed special wisdom to guide our food choices.

THE PROBLEM WITH PATERNALISM

Although karma is a concept often associated with Eastern mysticism, it is actually an idea deeply embedded in the thinking of most Americans. We have a fundamental belief that people should get what's coming to them. A key factor driving anger at Wall Street in the most recent financial disaster was the bank bailout. The entities that many blame for the cause of the financial meltdown were the very recipients of a government handout. The banks' actions seemed to justify bankruptcy and CEO firing, but instead we saw bailouts and bonuses. Such actions violated a fundamental sense of fairness among many Americans: it violated karma.

The emergence of behavioral economics and the food police has led to at least one unintended consequence: it has fostered a culture in which there is an implicit abdication of personal responsibility. Behavioral economics, at its best, serves to remind us that we are not perfect. But it has also helped normalize abnormality. By arguing that we lack the ability to choose wisely, some behavioral economists have implicitly promoted the idea that we therefore cannot be judged for making poor choices. We do not punish a toddler the first time he spills milk, because we understand that he lacks the ability to control his actions or to comprehend the consequences of his choices. Likewise, behavioral economics research promotes a climate in which it seems inappropriate to let people suffer the consequences of their poor choices because it suggests that those people are too irrational or unknowledgeable to do better. When the food police use behavioral economics to promote everything from fat taxes to bans on restaurant advertising, the notion that people should not be held accountable for their actions becomes institutionalized and reduces the incentive for our taking personal responsibility.

If people believe their government will step in and solve problems that arise from decisions to save too little or eat unhealthily, then there is little incentive to alter current bad behavior. It has been argued that one of the many problems that led to the 2008 financial crisis was that banks knew they would be bailed out, even if they made poor decisions and took unwise risks.[28] Likewise, no matter how caring they seem (and in some cases are), generous government-funded retirement plans and health care serve to bail out people who chose not to consider the future and, in the process, to reduce our incentives to take care of our future selves. We are smart enough to know not to save for retirement if someone else does it for us.

I am not arguing that the poor and elderly should be ignored. Concerns for equity can legitimately motivate actions to help. Nevertheless, we move into dangerous territory when such actions are justified on the grounds of cognitive failures. Observing that people act foolishly in some circumstances does not excuse said behavior. Food police who seek to rescue us from our own fallibilities will, in the process, compound the problem by diminishing our incentives to rescue ourselves.

Brian Wansink's book *Mindless Eating* points out the myriad ways our food choices are influenced by small and irrelevant details. But, interestingly, Wansink doesn't call for government action to limit plate sizes or ban buffet lines. Rather, he recognizes the power of individuals to make use of behavioral economics research to better their own lives. He promises to show "how to remove the cues that cause you to overeat and how to reengineer your kitchen and your habits . . . You can eat too much without knowing it, but you can also eat less without knowing it."[29] We don't need the government to nudge us. We can nudge ourselves. The power to nudge should be left in the hands of those who know when the nudge is needed.

WHAT'S THE ALTERNATIVE?

We are often told that there is no alternative to paternalism; that the government and others must make some decision that will influence our choices. To make their point, Thaler and Sunstein ask us to put ourselves in the role of a cafeteria manager. "Consider the problem facing the director of a company cafeteria who discovers that the order in which food is arranged influences the choices people make. To simplify, consider three alternative strategies: (1) she could make choices that she thinks would make the customers best off; (2) she could make choices at random; or (3) she could maliciously choose those items that she thinks would make the customers as obese as possible."[30] What should we do?

The first thing we should do is reject the premise of the question. There is an alternative noticeably absent from the list: arrange the foods in a way to maximize profit.[31] Profit seeking often has a negative connotation, but a surefire way for firms to make profits is to give consumers what they want. Detractors of profit seeking falsely envision an evil monopolist at work, but the reality is that firms are in fierce competition for our business—especially restaurants, which routinely go bust when they cannot supply what the customer wants at a price he is willing to pay. Thus, in a sense, the cafeteria manager has very little control over where to arrange the desserts;

if the manager wants to stay in business, she will place the desserts in a location that is most likely to induce consumers to buy them.

Even if the behavioral economists are right and the consumers do not actually know what they want, enterprising cafeteria managers will try out different combinations of food arrangements until they find the one that yields the most money. Their competitors will simultaneously try the same, each using whatever means is at his disposal to convince (or nudge) people that his arrangement is best. This dynamic competition embodied in the marketplace, predicated on voluntary exchanges between willing buyers and sellers, can be said to "give people what they want and are willing to pay for, when they want it and are willing to pay for it."[32]

We may change our minds tomorrow about what we want, but an energetic marketplace will seek to anticipate our mood swings and satisfy our changing preferences. And even if our preferences are incoherent in a way that makes traditional economic welfare analysis nonsensical, we can take joy living in a world that attempts to anticipate our changing needs and presents us with a variety of options. Freedom and choice have intrinsic value. As economist Robert Sugden put it, "[B]eing able to choose how to live one's life is an aspect of individual well being in its own right."[33] Nobel Prize–winning economist Amartya Sen has argued, "Choosing may itself be a valuable part of living, and a life of genuine choice with serious options may be seen to be—for that reason—richer."[34]

Government bureaucracy lacks this key dynamism embedded in the market. A paternalist seeking to improve our health might mandate that all cafeterias place the fruit and vegetables first in line, based on research showing that this will increase those items' uptake. But what will ensure that this change won't cause people to eat at McDonald's instead or bring their lunch from home or eat more snacks later, or even cause the cafeteria to go out of business? If McDonald's truly is the evil empire the food police claim it to be, why wouldn't McDonald's pressure the rule makers to get the outcomes they wanted? *The Economist* criticized Thaler and Sunstein's proposal to give the government the power to nudge by saying, "There is a serious danger of overreach . . . Politicians, after all, are hardly strangers to the art of framing the public's choices and rigging decisions for partisan ends. And what is to stop lobbyists, axe-grinders and busybodies of all kinds hijacking the whole effort?"[35] The food police can rest satisfied in nudging us toward greater health, all the while being unaffected by complex interactions that lead to outcomes the paternalist never intended. The same cannot be said of the cafeteria owner who puts her wallet on the line every time she reorganizes the buffet line.

The economist Robert Sugden says the paternalists are wrong: "There *is* a viable alternative to paternalism. It is what it always has been—the market."[36] I don't know about you, but when I walk into a cafeteria, I *want* the manager to tempt me—to offer me what looks good at a price I'm willing to pay. Yes, this kind of world might require more willpower and self-control, but the alternative world of the paternalistic food police would deprive us of the noble act of making a wise choice when we had the freedom to do otherwise.

THE FASHION FOOD POLICE

ORGANIC—THE STATUS FOOD

According to ancient Jewish Scripture, man's first moral quandary began with a choice of whether to eat a piece of fruit. Despite the passage of millennia, we find ourselves back in Eve's proverbial fig leaves. There is the inexpensive Red Delicious, but who'd be caught buying something so boring? Express some individualism and go for the Granny Smith, or push the basket down a bit farther, prices rising all the way, and go for the organic Gala. Even though budgets are tight, there's no telling what has been sprayed on the industrially grown apples, and after all, we should consider the health of our children. Or at least consider the kids' teachers, who would think us the worst kind of parent and environmental deviant if we didn't double the shopping bill for organic. After all, an adviser for *Consumer Reports* magazine tells us that nonorganic farmers are "raping the land."[1] Better yet, hop in the SUV and save the environment by heading down to the farmers' market. Of course, we'll turn a blind eye to the carbon footprint of the Cadillac Escalade. Unlike Eve, at least we can take some comfort in knowing that the fate of humanity doesn't hinge on our apple choice. Or does it?

Let's face it: food choice was once much easier, though much less exciting. Today's supermarkets boast over a hundred types of breakfast cereal alone. Which will you pick? Are you a Fruit Loop or 100 percent organic granola? Both will fill your belly, but only one will earn the approving nod of the fashion food police. Their message is clear: if you don't choose organic, not only are you unhip, but you've sided with the serpent.

In one sense, the rise of organic food corresponds with the libertarian ideal. On the one side were consumers who demanded new attributes from the food system, and meeting them were entrepreneurial farmers willing to change production practices to make an extra buck. There's only one thing wrong with this organic-is-libertarian story. It isn't true. Consider Julia Roberts's assertion on the *Oprah Winfrey Show* that it's our responsibility to be green for our children and eat organic food. Julia's green food guru, Sophie Uliano, tells listeners that if they can't find organic items in their local supermarket, "Speak up! . . . We as women, we are powerful. We have a voice . . . go to the manager and say to him, 'I want to see more organic here!' "[2] Ah, sounds just like un-coerced choices in a free market. As if store managers need to be yelled at to offer products that will bring more profit.

Hordes of consumers are being scared, mothered, and guilted into buying organic, and political pressures are redirecting millions of taxpayer dollars to support organic production. We're often told that organics don't get government subsidies, but that's a fabrication. In Europe, organic farmers are subsidized like all other farmers. In the United States there are programs such as the Environmental Quality Incentives Program, which pays producers to transition from conventional to organic. Other programs use federal monies to help organic farmers pay the cost of certification. Organic farmers can receive government-subsidized crop insurance just like nonorganic farmers. Organic milk is subjected to many of the same complex price-support rules imposed by the government on nonorganic milk.[3] Hundreds of millions of tax dollars are spent on research and education into organics and on marketing and monitoring programs.[4] The food police tell us that the growth in organic food demand is a result of the free market working at its best, even as they use the taxing power of the state to manipulate the market by subsidizing organic production, marketing, and research activities. You can't have your organic-is-libertarian cake and eat it, too.

When we spend a few extra bucks for the shirt adorned with a horse or alligator, most of us know what we're getting. We're paying a few extra bucks to be stylish, to fit in, to follow a trend. Deep down, if we're honest with ourselves, we know the cheaper, logo-free shirts aren't really much different from their stylish cousins. The trouble with organics is that people believe the hype. By all means, if you want to buy organic to be stylish or look good to your neighbors, then go for it—it might be a small price to pay to fit in, depending on your income. But let's stop with the false advertising.

Turn on almost any daytime talk show or watch any

celebrity chef, and you'll hear the same drivel. On Oprah's show, we hear "Organic farming protects the planet, so it's a win-win. It's healthier for us, and it's healthier for the planet."[5] In *Whole Living* magazine, the editors tell us they think organic strawberries and carrots taste sweeter, and we are encouraged to "Buy organic whenever you can."[6]

I'm truly sympathetic to those who want to buy healthier food or who are trying to protect their families from the negative side effects of pesticides. I have a family, too. But don't we also want to make sure our money is well spent? After all, the kids have to go to college; we have the bills for art lessons and sports; safer automobiles and car seats are expensive, too. Money spent ratcheting up our shopping bill for organic food is money not spent on other things we also value for our families. One can literally spend thousands of extra dollars a year buying organic rather than conventional.[7] That's thousands of dollars *not* given to a charity supporting the homeless, saved for retirement, or available for Junior's summer camp. We all want our children to grow up healthy, but we have to marry our concerns with the evidence on whether the actions we're taking have any impact.

One of the problems with organic is that few shoppers know what it really means, and they project onto the nebulous word all their hopes and dreams of eating in a safe, healthy, environmentally friendly manner. *Organic* is like the word *natural*. Whatever you may think *natural* means, the USDA primarily defines it as "minimally processed." Under this legal definition, almost any beefsteak, pork chop, or chicken breast can be labeled "natural." Likewise, the USDA precisely defines the word *organic*. Farmers can use the label if they grow crops and livestock according to the list of criteria established by the government. In essence, organic crops are grown without synthetic pesticides or fertilizers, and farmers must adhere to certain tillage and cultivation practices. Producers of organic animal products must feed livestock organic grain, allow them outdoor access, and refrain from administering them added growth hormones or antibiotics.

"Organic" means food produced according to a set of rules. But in no way do the rules guarantee that organic is safer, tastier, healthier, sustainable, or more environmentally friendly. It is, in fact, so hard to tell organic and conventional apart that a dozen Italian companies were recently caught selling 700,000 tons of counterfeit organics all over Europe.[8] The organic label doesn't mean the food came from small farms, is without pesticides, or wasn't processed by a large agribusiness. Indeed, the supposed job-destroying, community-crushing villain Walmart has cashed in on the craze and is now the largest seller of organic food in America.[9] It is time to set the facts straight on organics.

Let's get one thing out of the way: there is absolutely no consistent scientific evidence that organic food is any tastier than nonorganic food. I know, I know. *You* can tell that organic food tastes better than that sleazy conventional stuff. I can't go around conducting blind taste tests with every organic food zealot I meet. But if you happen to find yourself in that crowd, I challenge you to find a friend (a neutral one without skin in the game) to set up a blind taste test to see if you really can tell the difference across several products. I'd bet good money you can't consistently pick the organic. Carefully controlled scientific study after study, published in peer-reviewed journals, show that most people, most of the time, just can't tell the difference.[10]

Sure, some foods labeled organic taste better than conventional foods, but it isn't the organic aspect that's causing the taste difference. Fresher fruits and veggies will almost always taste better than non-fresh—organic or not. Some crop varieties are tastier than others regardless of whether they're grown conventional or organic. Different amounts of rainfall, wind, and any number of complicating factors can affect the taste of food. But don't be fooled into believing that organic per se has any consistently measurable effect on the taste of food.

Let me tell you another shocking truth: many people believe organic tastes better than conventional because our expectations are often more powerful than our taste buds. Research has shown that people will say the same bottle of wine tastes better if it has a higher price tag.[11] Brain scan studies show that most people can't tell the difference between Coke and Pepsi in a blind taste test, but when told they're drinking Coke, their brains suddenly get fired up.[12] Consumers similarly tend to think organic will taste better than it actually does. They say the organic option tastes better—but only *after* they've learned it is organic.[13] So, yes, organic food might really taste better to you. But it's your brain telling your tongue what to think, not the other way around.

While we're on the topic, there are a lot of things associated with organic that do affect the taste of food, and sometimes for the worse. There is the myth about organic milk being naturally safer than regular milk because it has a longer expiration date. Organic milk often has a longer shelf life not because it is organic but because it has been subjected to ultra-high-temperature pasteurization.[14] Dairies use the technology because organic milk often has to travel farther and sit on the shelf longer due to the lower volume of the product purchased due to its higher prices. But most people don't like the taste of milk that's been ultra-high-temperature pasteurized, which is why conventional dairies in the United States don't use the process. I spent much of the last year in France, and most of the milk there was sold right off the unrefrigerated shelf—next to the laundry detergent. Why? Because it is ultra-high-temperature pasteurized. And you want to know something? It tastes terrible.[15]

Let's get to a more serious matter. Is it healthier to eat organic food than nonorganic? Some studies claim organics are more nutritional; others show the opposite. In cases like this it is useful to look at large-scale literature reviews such as the one commissioned by the British Food Standards Agency and published in the *American Journal of Clinical Nutrition,* which found, after sorting through thousands of studies, that "there is no evidence of a difference in nutrient quality between organically and conventionally produced foodstuffs."[16] Not surprisingly, the organic industry has contested these findings—but what would you expect from a group with such a vested interest in the outcome? The fact that one or two studies (out of more than a hundred studying the issue) can be turned up showing a higher level of one nutrient (out of a dozen nutrients tested) in organic than conventional only serves to show that any "true" relationship between nutritional content and organic that might exist is exceedingly small and probably within the margin of error scientists are able to consistently detect. That's hardly a ringing endorsement for the supposed added healthfulness of organic.

There is one troubling facet to the dubious "organic is healthier" claim touted by the food police. Because of the exalted status ascribed to organic, the research shows that people are readily fooled into thinking organic foods are much healthier than they actually are. Even when organic and conventional foods are labeled with the same caloric content, people believe the organics are somehow magically lower in calories. And if tasked to lose weight, apparently people believe it is more acceptable to give up exercise if they simply eat organic rather than conventional. These sorts of findings led one team of psychologists to conclude that organic is "biasing everyday judgments about diet and exercise."[17] No wonder the following exchange was overheard in the checkout lane of a natural foods store:

"Mom, look! Organic gummy bears!"

"Yes, I see. No more sweets."

"Mom, but they're organic."[18]

But surely organics are safer because they don't contain pesticides, right? The case is less obvious than it first appears. First, it is a complete misconception that the cultivation of organic food does not involve pesticides. Organic farmers are free to use "natural," or nonsynthetic, pesticide, herbicide, and fertilizer. The trouble is that many of these "natural" pesticides are just as toxic and carcinogenic as their synthetic counterparts. For example, various "natural" yet toxic copper compounds are routinely used as fungicides, and "natural" sulfur is routinely used as a pesticide in organic agriculture. Moreover, many people fail to realize that plants naturally make their own pesticides. These self-made pesticides protect the plant but are dangerous to us. In fact, 99.9 percent of the pesticides we consume are natural, while only 0.01 percent are synthetic. Many of these naturally grown pesticides are even deadlier than the synthetic ones we humans have devised.[19]

In general (but not always), organic agriculture tends to use a lower total volume of added pesticides and herbicides than conventional agriculture, but there is really no way to tell for sure whether that organic apple you're paying a premium for has more or fewer chemicals than the much cheaper alternative. One summary in the *Journal of Food Science* stated, "It is premature to conclude that either food system [organic or conventional] is superior to the other with respect to safety or nutritional composition. Pesticide residues, naturally occurring toxins, nitrates, and polyphenolic compounds exert their health risks or benefits on a dose-related basis, and data do not yet exist to ascertain whether the differences in the levels of such chemicals between organic foods and conventional foods are of biological significance."[20]

As far as the synthetic pesticides on conventional fruits and veggies go, the government already places limits on the amount and type of residues allowed, and it uses a variety of scientific litmus tests to determine residue levels that are deemed safe. Moreover, simply washing fruits and vegetables can remove at least some of the pesticide residue. In the end, there is *some* legitimacy to the pesticide-reduction claims of organic, but the facts about the safety benefits are more equivocal than most realize.

So, you *might* be able to reduce risks from pesticides by eating organics, but how risky are these chemicals anyway?

Synthetic pesticides in food are a relatively small risk in the grand scheme of things. To put it in perspective, consider the fact that many foods we routinely eat contain large amounts of natural compounds that are far more toxic than the pesticides that farmers spray on them. It has been estimated that "three daily cups of coffee or one gram of basil a day is more than 60 times as risky as the most toxic pesticide at current levels of intake."[21] All foods have naturally existing chemicals, and the research shows that the natural substances in foods such as coffee, basil, lettuce, mushrooms, orange juice, and many others are more carcinogenic than the typical amount of synthetic pesticide residue remaining from farmers' attempts to ward off bugs.

If your choice is between eating conventional fruits and veggies sprayed with pesticides or eating none at all (perhaps because organic is too expensive), you are *much* better off eating the produce grown with synthetic pesticides. That ought to tell you something about the magnitude of the risk in question. To illustrate, some studies have estimated that using pesticides in food production causes about twenty deaths per year in the United States. Twenty deaths are a tragedy. What about twenty-six *thousand* deaths? That spectacularly larger tragedy is the projected effect of a ban on pesticides, which would drive up prices of fruits and vegetables (and thus drive down their consumption and increase cancer rates).[22] Eating fruits and vegetables is much healthier than eating pesticides is risky.

A complete phaseout of pesticides is projected to cost U.S. consumers and agricultural producers at least $20 billion per year. If we spent $20 billion saving twenty lives by eliminating pesticides, this means $20 billion less to spend on other things—such as tests for radon, which are projected to save about fifteen thousand lives for less than $20 billion. If the choice is to spend $20 billion saving twenty lives or spend the same amount saving fifteen thousand lives, I know how I will vote. In short, pesticide use in food is a legitimate worry, but let's make sure our worries are properly prioritized.

There is more to food safety than synthetic pesticide use. Said one former FDA official, "Here is a misperception, I think with many people, that organic equals safer, and it's just not. It's dangerous, actually, because it creates a false sense of security."[23] The danger to which the official was referring was the presence of *E. coli* in organic spinach, which sickened about 275 people and caused 3 deaths back in 2006. In 2011, organic sprouts killed almost 40 Europeans and sickened at least 3,800 more. That same year, First Class Foods recalled 34,373 pounds of organic beef. These are just a few examples of organic food safety recalls.[24]

There are aspects of organic production that might make organic food more susceptible to bacterial contamination than conventionally grown food. Unable to use synthetic nitrogen fertilizer, organic growers often use animal manure, which, if not properly composted and managed, can contaminate fruits and vegetables with *E. coli*. One prominent study showed that the percentage of positive samples of *E. coli* was more than six times higher in organic than nonorganic produce (though prevalence rates for organic and nonorganic were similar). Some organic produce on farms with poor manure-management practices had a prevalence of *E. coli* nineteen times greater than that on other farms. The study's authors concluded, "The results of the present study do not support allegations that organic produce poses a substantially greater risk of pathogen contamination than does conventional produce. However, the observation that the prevalence of *E. coli* was significantly higher in organic produce supports the idea that organic produce is more susceptible to fecal contamination."[25] The jury is still out on the extent to which organics are generally more (or perhaps less) susceptible to bacterial contamination. However, the available evidence suggests that organic foods appear no less likely to be contaminated by *E. coli* or *Salmonella* than conventional foods.

As I see it, the noteworthy benefit of organic relates to environmental outcomes, but here, too, the purported advantages, while real, are overhyped. Organic producers often use no-till production systems (in which farmers do not plow the ground as often) and use cover crops that reduce the need for irrigation and help prevent soil runoff. No-till production is a real environmental benefit, but that's why many *non*organic producers already use it. For example, nationwide, more than 35 percent of U.S. cropland is no-till.[26] One recent survey in Oklahoma showed that only 38 percent of wheat acreage is farmed using conventional tillage practices. The remaining 62 percent is farmed using some form of minimal-till or no-till production.[27] So, a key environmental benefit of organics, reduced or no tillage, is one that many nonorganic producers are already enjoying. The point: you don't have to double your shopping bill to get the benefits of reduced land tillage.

One key problem with organics is that their yields are lower. A paper published in the prestigious journal *Science*, for example, reported 20 percent lower crop yields on organic farms.[28] A USDA study showed that organic dairies produce 30 percent less milk per cow than nonorganic dairies.[29] Data from a large-scale survey by the USDA imply that yields were 35 percent lower for fruits, nuts, and berries; 30 percent lower for field crops; and 38 percent lower for vegetables on organic farms as compared with nonorganic farms.[30]

Many organic advocates like to dispute the yield-reducing effect of organic. However, if organics could actually increase yield and reduce input use, as organic advocates like to claim, what farmer in his right mind wouldn't switch? Better yields, higher prices, lower input costs: organics should be a no-brainer. Yet less than 1 percent of all U.S. cropland is organic.[31] Either farmers are *really* stupid or the purported yield benefits are a myth. The truth is that organics yield significantly lower amounts of the food people want to eat per acre planted. Thus, to produce the same amount of food we currently enjoy, organic would require much more land. Switching to organic would likely mean bringing into production land that now sits idle. That hardly sounds like an environmental benefit to me.

The shouting match between organic proponents and detractors is a bit nauseating: show me your published study and I'll show you mine. So, let's put all the scientific studies aside and look at an incontrovertible fact facing us in the grocery store: organics sell at 40 percent to 100 percent to even 200 percent premiums to nonorganics. Economics teaches us that price differences are important though sometimes imperfect signals regarding resource use. The price we pay for food in the grocery store must reflect the costs that went into producing that food: from land rent to the value of the farmer's labor to the prices of seed and fertilizer. Higher prices for organic means that, somewhere along the line, organics used more land, more labor, more seed, more fertilizer, or more of any of the other inputs required to produce food. The prices of all these inputs were each determined by their scarcity relative to people's desires to use them for other purposes unrelated to food production. So, when we see that organic is higher priced, it signals to us that organics are using many more of the resources society finds valuable than are nonorganics. Using up more resources is exactly the opposite of "sustainable."

Normally the price mechanism is used to ration scarce resources and to show us how to allocate resources over time. Rising prices for increasingly scarce resources such as oil and fertilizer cause us naturally to back away from consuming those resources—thereby resulting in a "sustainable" future. The fact that we are now using a lot of oil and fertilizer in agriculture means these resources are currently in ample supply relative to demand. When the prices of oil and fertilizer rise, we'll use less of them. The sustainability movement largely represents an elitist attempt to ration scarce resources using social pressure, guilt, and regulation.

If the benefits of organic are really so small and the costs so high, why all the fuss? Organics are the ultimate example of a culture that has turned eating into a status-seeking enterprise. Martha Stewart recently remarked, "Food has become more than one of life's great pleasures. It has become a signifier of style, too . . . These days, it often seems that you are what you purchase in the supermarket or at the farmer's market; your grocery list is a reflection of your values and your identity."[32]

Not only does she hit the nail on the head; she's driving the hammer.

In our modern world of variety and abundance, eating should be a pleasure, and not a *guilty* one. It used to be said that you are what you eat, a motto that stressed individual choice and responsibility. But now it seems that what you eat conveys who you are. What we eat, where we shop, and how we cook have become symbols for who we are as individuals. The modern consumer is forced into a search for meaning in her daily food choices. We are no longer able to define ourselves separately from what we eat. Eating is no longer a decision about how to fill our bellies; it has been made out to be a symbolic choice with moralistic overtones.

The food police have pronounced eating at McDonald's a sin, not because there are tastier or healthier alternatives elsewhere but because eating at McDonald's is a fashion faux pas. It would be unthinkable to question someone's moral character because they chose not to drive a Mercedes or carry the latest Louis Vuitton bag. But when it comes to food, those who cannot display the appropriate level of status through what they eat are pitied and vilified. Martha Stewart doesn't pull any punches when she reveals, "What's in your pantry and on your plate have become a form of self-expression much like a fabulous pair of Christian Louboutins, or absolutely anything vintage."[33] Now that the progressive food elite have defined the morally acceptable food, it is said that it is unjust that the poor cannot afford to live up to the standard. A *Newsweek* article recently opined that "[a]s more of us indulge our passion for local, organic delicacies, a growing number of Americans don't have enough nutritious food to eat."[34]

Not only are the poor unable to keep up with the foodie Joneses, but the food police apparently think the poor are too dumb to realize what's best for their pocketbooks. *New York Times* columnist Mark Bittman says "junk food" isn't actually cheaper than what he calls "real food."[35] In a back-of-the-envelope calculation that is supposed to be taken for serious scholarship, he figures that a family of four can eat roasted chicken and vegetables at home more cheaply than dining at McDonald's.

His comparison is utterly flawed because he fails to count the time costs of cooking and cleaning. Instead, Bittman calls cooking a virtue. Tell that to the single mom working sixty hours a week! It is of course possible to spend less money by eating at home, even factoring in time costs, but did Bittman ever stop to think some people eat at McDonald's not because it's cheaper but because they want to? It was Adam Smith who reportedly noted that "[v]irtue is more to be feared than vice, because its excesses are not subject to the regulation of conscience."

The paradox of the food police is that they have turned food into a status-seeking game while simultaneously asking why the poor don't have enough nutritious food to eat. It's like the rich kids at school honestly wondering why everyone isn't wearing the most fashionable jeans.

Status seeking and shame are two of the primary tools used by the food police. Just ask your kids. Apparently oblivious to the need to ensure competence in reading, writing, and arithmetic, many schoolteachers and administrators find time in the day to preach about local foods, social justice, and recycling. One mother interviewed by the *New York Times* about schools' attempts to go green reported that "the social pressure her children felt regarding recyclable products was palpable." Her sin was to use plastic bags in the lunchbox, an infraction so heinous as to invoke a teacher's wrath. Said the mom, "That's when the kids have meltdowns, because they don't want to be shamed at school."[36]

The progressive food movement will not meaningfully help the poor or the environment or public health: it is a way for a modern generation far removed from the farm to give meaning to their lives in how they define themselves and others through food. It is a movement that has turned food choice into a status-seeking enterprise, shaming people who will not play the game. As a result, platitudes are mistaken for pragmatic solutions, and hypothetical survey responses are interpreted as signs of market failure. Now that the food police have pronounced what is trendy and have made the media rounds to persuade the masses, a culture has been created in which people will say and believe all kinds of things to fit in.

The truth is that we often bloviate in dinner groups or give responses on surveys because we want to look good to the audience, never stopping to consider our own actions. We even want to look good to ourselves—and why not give yourself the chance to feel good by saying you support organic or local foods in a survey, especially when you don't really have to pay the cost? It leads to the same kind of hypocrisy (or perhaps just plain elitism) illustrated by Al Gore spewing carbon out the back of his jet traversing the country telling the rest of us about his inconvenient truths.

I saw the social pressure to buy fashionable food firsthand a couple years ago when a colleague and I conducted a study asking consumers whether they'd like to buy two grocery items: environmentally friendly dishwashing liquid and locally grown organic ground beef. We asked the study participants if they wanted the new products or the more traditional brand-name products such as Dawn and Palmolive. Some people answered the questions hypothetically, and others actually had to put their money where their mouth was.[37]

Then came the vital stage of the study. We actually put the trendy dishwashing liquid and beef for sale in a local grocery store. The big difference was that the shoppers had no idea these new products were part of our research study. They thought they were out for a routine shopping trip, unaware we were watching what they chose.

So, how did our recruited participants, who knew they were taking part in a research study, compare with regular shoppers who had no clue that their decisions were being scrutinized? The choices made by the participants who knew we were watching them led us to predict that the store would sell 386 pounds of the organic ground beef in a month's time. In reality, only 12 pounds were sold in the store—32 times less than predicted! Apparently, people told us they liked organic beef much more than they actually do. Likewise, the participants who knew they were being studied chose the environmentally friendly dishwashing liquid 25 percent of the time over Joy, Dawn, and Palmolive. But not a single bottle of the fashionable soap was sold in the store! It seems people want to project the image of being the kind of people who buy environmentally friendly products—*but only if they know someone is watching.*

We want to *appear* to be the kind of people who buy organic and environmentally friendly foods. But deep down, it's not really about organics or costs and benefits. It's about keeping up with the Joneses. Despite what we *say,* many of us buy organic not because we think it improves the environment or our health, but because it makes us *feel good* about the kind of people we tell ourselves we are. What this means is that we undertake a lot of potentially wasteful activities just to fit in.

This kind of status-seeking consumerism or conspicuous consumption, as the economists call it, arises when we buy things simply to improve our relative social standing. My Ford Taurus is likely to seem just fine—until my neighbor pulls up in a Cadillac. No doubt a new Lexus will be in order. Rather than curtail this kind of behavior, the food police have promoted it.

At the end of the day, choosing whether to buy organics is a tough trade-off. Organics are neither tastier nor healthier than nonorganics. While organics tend to use lower amounts of pesticide, they are not free of pesticide use, and the health risks from pesticides rank pretty low relative to the other risks we face in life. Organic production *might* be relatively better for the environment depending on yield response and the extent of conservation tillage practices conventional farmers already use. One thing is sure: organics cost much more. This is no easy choice, and I won't disparage anyone for whichever apple I find in their cart. All I'm asking is that the food police do the same.

6

FRANKEN-FEARS

I clearly remember the first time I saw a genetically modified plant. I was a freshman in college working for an agricultural extension agent who had planted some demonstration plots for local farmers. I had never before heard of biotechnology or genetically modified organisms (GMOs), but when I saw the luscious green rows of herbicide-resistant cotton standing next to the browned and wilted conventional variety, I was overjoyed. The reason? It meant the end to a job I had grown to hate: walking miles and miles in the sweltering sun earning calloused palms hoeing cotton weeds. A single tractor ride through the field and a squirt of herbicide could now put an end to my misery. Apparently the farmers were equally impressed, because within ten years virtually all the nation's cotton crop would be genetically modified.

Only later did I learn that the same crop that brought me

such delight was greeted with shock and horror by the food police. These are the same food police who tell us they're worried about the environment and our health but who ignore the science showing that genetically modified crops can improve both. Many of these same people call any restriction on stem cell research an affront to science and reason; and yet, if the same research is used to create faster-growing wheat that reduces the price of bread for consumers, it is deemed an unholy revolt against nature.

Local food activist and celebrity chef Alice Waters not only outlawed biotech foods from her own restaurant, but she also persuaded the Berkeley, California, public schools to do the same. However, she might want to double-check the grilled Paine Farm squab with Chino Ranch peppers and sweet corn fritters that was recently listed on the menu of her restaurant Chez Panisse. You see, the genetics of the corn used in those fritters are nothing like what Mother Nature originally created. About ten thousand years ago, the ancestor of a modern corn cob was no longer than your thumb. With our forefathers and -mothers repeatedly picking and planting only those cobs that were tasty and big enough to merit the effort, we eventually ended up with corn cobs that today average about a foot long. So, when you hear you're eating natural food at Chez Panisse, you might ask which century of nature they're talking about. Like it or not, our ancestors genetically modified corn. Perhaps they didn't do it with the tools we now have, but make no mistake about it, we've been altering genes since Adam spat the first seed on the ground.

The fascination with "natural" food bears almost no relationship to the historical reality of food. Ten thousand years ago, wild rice was little more than a stalk of grass, which Asian hunter-gathèrers had to pounce on before the edible seeds shattered or blew away with the wind.[1] Only by interfering with Mother Nature did we reach the point where rice can now account for one fifth of the world's total caloric intake. The modern potato contains a small amount of natural toxins that are mostly harmless, but it reached its current edible state only after centuries of selective breeding. The first "natural" potatoes were more toxic and often required ancient Peruvians to lay the spuds on the cold ground overnight to help remove the toxicity.[2] How's that for ancient freeze-drying?

Rather than trying to debate the facts, the food police have resorted to scare tactics and name-calling. We are told that biotech products are Franken-foods. But watch one of the Frankenstein movies and see who's scarier: the artificial monster or the foaming-at-the-mouth mob of protesters with pitchforks. Greenpeace, Friends of the Earth, the Sierra Club, and a host of other activist groups want us to imagine Mary Shelley's beast and pick up our torches upon mention of the technology, but the reality is something closer to Herman Munster from the 1960s sitcom: an amiable creature who can do a lot of good in the world if only properly understood.

We've been told for more than a decade that biotech crops will bring about environmental Armageddon. None other than Britain's Prince Charles, being the eminent scientist that he is, has warned that biotechnology "will be guaranteed to cause the biggest disaster environmentally of all time."[3] But we've been planting the stuff for almost two decades now. In 2011 about 90 percent of all corn, cotton, and soybean acres in the United States were planted with genetically modified seeds.[4] Where is the catastrophe? Perhaps there's a better question: Where is the accountability for all those who wrongly prophesied gloom and doom?

Rather, when we look at the scientific research, a much different story emerges. USDA studies show that farmers who plant insect-resistant biotech corn and soybeans reduce insecticide use. Moreover, the adoption of herbicide-resistant soybeans is associated with increased use of conservation tillage, which helps prevent soil erosion and reduces the need for irrigation.[5] A careful study reported in the prestigious journal *Science* showed that in rural India, the growing of biotech cotton reduced insecticide use by almost 70 percent while simultaneously increasing yield by more than 80 percent.[6] So, yes, while there are *potential* environmental risks, there are *demonstrably real* environmental benefits from using biotechnology.

The National Research Council of the U.S. National Academies recently concluded, "Many U.S. farmers who grow genetically engineered (GE) crops are realizing substantial economic and environmental benefits—such as lower production costs, fewer pest problems, reduced use of pesticides, and better yields."[7] After all the hype and misinformation about dying Monarch butterflies and the dire warnings of toxicity and allergenic effects, and after more than a decade of our eating biotech foods, there has not been a single scientifically confirmed case of human illness that can be attributed to food biotechnology. Claims that current biotech foods are less safe, less nutritional, and less tasty than their non-biotech counterparts have absolutely no grounding in scientific evidence, according to virtually every major scientific authority on the subject.[8]

Let me be clear. I am not necessarily trying to convince you to eat genetically modified food. I am personally convinced that the current biotech products on the market are safe and are net-plus for the environment. But you don't have to agree with me. If you don't want to eat products made with biotechnology, then buy the organic products certified to be GMO free. "But organic is much more expensive," you say. Yes, that's because it is more expensive to produce food without biotechnology. You don't have to like my decision, but don't ask me to subsidize yours—and have the courage to let others arrive at a different conclusion from yours.

It is true that, when asked, most people say they prefer to avoid foods made with biotechnology, at least the herbicide-resistant corn and soybean varieties. But my research clearly shows that there are many people who would be willing to switch for a small price discount. Even in a country such as France, which is supposedly vehemently anti-GMO, research shows that 23 percent of consumers are indifferent to foods made with biotech and another 42 percent are willing to purchase GMOs if they are sufficiently inexpensive. So, when the food police decide to limit and ban applications of food biotechnology, remember that they're making a choice for others that many would not have made for themselves.

After all, nobody's saying biotechnology shouldn't be regulated. And regulated it is. New biotech crops have to go through three regulatory agencies: the USDA, the EPA, *and* the FDA. There have been reams of reviews by groups such as the UN Food and Agriculture Organization, the U.S. National Research Council, and the World Health Organization. Farmers who want to plant insect-resistant biotech corn must also plant "refuges" without the technology (refuges are small patches of nonbiotech corn planted around the larger fields of biotech corn), to mitigate against development of insect resistance. Insect resistance, by the way, is a problem with *all* types of insecticides, whether "natural," synthetic, or biotech. Companies introducing new food biotech products have to submit safety evaluations to the FDA for approval and must establish "substantial equivalence" between the new product and the unmodified variety (which basically means they must show that the new, modified variety is exactly the same as the old variety in every way except for the new modification).

I've never quite understood the argument that biotech crops are underregulated. Rather, it seems that there are those who'd stop at nothing to prevent biotech crops from being planted and who wish they'd rigged the process to yield a different outcome. A common refrain is that companies are not required to prove there are no "long-term" risks, but it is never specified exactly what such critics mean by "long-term"—a target in constant motion. There's also the common complaint that biotech companies don't have to prove the new products won't cause illness. Logic dictates that you can't prove a negative. Alas, logic has never been a strong suit of the food police.

Rather, the food police are constantly calling for a ratcheting up of regulatory standards for approval of biotech products. All the while, they fail to realize their efforts serve only to strengthen their most hated adversary: Monsanto. Do an Internet search for Monsanto—the maker of the herbicide Roundup and the producer of most of the genetically modified seeds used in the United States—and the first adjective Google will suggest is the word *evil*. Such has been the influence of the food police that the most commonly coupled search term with Monsanto is the descriptor normally reserved for Satan himself. But who benefits from stricter regulations that make it harder for new biotech seeds to enter the market? It certainly isn't the small start-up firms trying to break down entry barriers to get their new invention on the market. Rather, it's the established behemoths who have teams of lawyers and lobbyists who can absorb the regulatory costs that keep out their smaller competitors. Adding regulatory hurdles hasn't dampened Monsanto's market power; it has enhanced it—another unintended consequence brought to you by the food police.

If you don't believe me, ask Ingo Potrykus, who created something called golden rice back in 1999. This modified rice produces beta carotene, which the body converts to vitamin A. Potrykus engineered the crop to help the millions of poor people around the world who risk becoming blind due to vitamin A deficiency. But despite all evidence that his product is safe, Potrykus has been unable to get golden rice approved internationally because of regulatory hurdles. How many other products that could help millions are collecting dust on the shelves of universities and start-up biotech companies because the regulatory costs are prohibitive? The expense of regulation keeps out small players such as Potrykus and helps convey market power to the established agrochemical firms.

If you didn't catch it, you read me right. Monsanto probably does have market power that conveys some control over the prices it sets for seeds and chemicals. So, yes, I know it's shocking: Monsanto makes money from selling seeds and chemicals. Documentaries such as *Food, Inc.* show us fearful farmers who claim to have been wrongfully sued by Monsanto for saving and accidentally planting genetically modified seeds rather than buying new seeds from the company. I have no idea about the accuracy of these claims, and I'm certainly not going to defend all of Monsanto's actions. It is true that Monsanto has sued some farmers (and in some cases won) for illegally planting the company's patented technologies.

But why the producers of *Food, Inc.* find the protection of property rights so abhorrent defies explanation. I suppose they won't press charges if I start selling bootlegged DVDs of their film out of the back of my truck at the farmers' market. Apparently Michael Pollan wouldn't have any problem with my photocopying *The Omnivore's Dilemma,* replacing his name with mine, and selling it for $.99 online. No, I suspect these folks would be rightfully indignant if their hard work and copyrights were so blatantly trampled. Why the double standard when it comes to Monsanto? Why are they to be faulted for protecting patents that took millions of dollars and decades of research to secure?

Just look at millions of acres in the United States planted with GMOs, and the reality is that the vast majority of farmers *willingly* decide to buy these seeds and plant them. The food police with no stake in the game presume to know more than each farmer looking at his bottom line and deciding what to buy. Does Monsanto charge more for its seeds than would be the case if there were more competition? Probably. But don't forget that Monsanto faces competition from rivals such as DuPont, Syngenta, Dow, and Bayer. If I could wave a magic wand, I would create more competition in the seed sector, but that doesn't overturn the obvious fact that despite the higher prices, farmers still find it in their interest to plant biotech seeds. I haven't yet read of Monsanto holding farmers at gunpoint, forcing them to adopt biotechnology—although it is amazing what conspiracy theories people will believe and circulate about Monsanto.

I've been in enough debates on this topic to know exactly what the anti-GMO foodie will say next: "Well, farmers don't really have a choice, because if they don't adopt biotech products, they can't stay in business." Bingo! That's how capitalism is supposed to work. The market doesn't reward us for producing what *we* want. It rewards us for supplying what *consumers* want. What's the alternative?

I'm sure there were a bunch of blacksmiths and wagon makers in 1908 who cast suspicious eyes on the new Model Ts being driven around town. Are we really to believe that the world would be a better place if only the wagon wheel makers got their way and avoided the supposedly corrupting influences of sleazy market forces? Wagon wheel makers had a choice, as farmers do now. They can provide what consumers want or they can find another line of work. And despite whatever we read in opinion polls, when we watch what most consumers actually do with their wallets when shopping, we see that what they want is lower prices.

Let's take a step back and ask how seriously can we take the claim that Monsanto is a drain on the system, sucking money away from farmers and consumers alike? The economic research shows all this to be nonsense. Farmers adopt biotech products because it makes them money and saves them time. Indeed, one would have to wonder if farmers were so stupid as not to understand their own interests, as virtually all major farm commodity organizations have lobbied on behalf of biotech. Consumers have benefited, too.

One study of the U.S. cotton industry found the biggest beneficiary of the advent of biotechnology was not Monsanto or the seed supplier, but U.S. farmers, who captured 59 percent of the benefits.[9] Another study, by the USDA, found that the development of herbicide-resistant soybeans benefited U.S. farmers (who captured 20 percent of the benefits) and U.S. consumers (who captured about 6.4 percent of the benefits).[10] Worldwide, the *annual* benefits from the creation and adoption of biotechnology are estimated at about $1.4 billion for cotton and $7 to $10 billion for soybeans and corn, and are projected to be more than $2 billion for rice. (No biotech rice is currently planted commercially.)[11] None of these estimates takes into account the potential benefits of the risk reduction that is provided by insect-resistant biotech seeds. Not only are yields and farm returns generally higher with biotech varieties, but they also save labor and serve as a form of insurance against some farm risks.[12]

Perhaps in our world of billion-dollar bailouts, some skeptic can claim these gains are too small to justify the potential risks. The truth is that it is really hard to know what we give up if we halt the advance of science. I've already mentioned genetically modified rice that can produce vitamin A, but scientists are also developing varieties of soybeans and corn that can make their own fertilizer and produce healthier cooking oil for humans; grasses that cause cows to produce less methane, a major greenhouse gas; drought-tolerant plant varieties suited for arid regions; tomatoes that can last longer without spoiling; bananas that make vaccines; strains of cassava with added micronutrients that could help millions of impoverished Africans who rely on the crop as a staple; and innumerable other products that have more tangible benefits for the consumer than perhaps the falling food bills we've already enjoyed. So, don't be fooled. Some biotech crops may be too risky and require prohibitions, but *not* seeing what benefits biotechnology has to offer may be the biggest risk of all. Just imagine all we would have missed had the first generation of Luddites decided that the world was "good enough" and halted all technological progress back in 1815.

Alas, we don't have to ask many of the poor farmers in Africa hypothetical questions about what they would give up. The food police have fought tooth and nail to keep the current technologies out of their hands. Content with their own food supply, many developed European countries are imposing their rich-country preferences on their poorer southern neighbors, and have thrown up roadblock after roadblock to biotechnology in Africa. As the political scientist Robert Paarlberg clearly shows in his book *Starved for Science,* published by Harvard University Press, organizations such as Greenpeace, Food First, and the International Federation of Organic Agriculture Movements have fought hard to keep biotechnology out of Africa.[13] While rich farmers and consumers in the United States are enjoying the benefits of insect-resistant corn, these groups somehow find it morally justifiable to hinder Africans from developing and using drought-tolerant maize.

Biotechnology is not a cure-all for Africa's woes, agricultural or otherwise, but for the food police to deny Africans a poverty-fighting tool that we already have is absurd. It's like telling a heart-attack victim he can't have an aspirin because he might get a tummy ache.

For those food police brave enough to travel to St. Louis and into the territory of the Evil Empire, they might want to take a look at what sits across the street from Monsanto: the nonprofit Donald Danforth Plant Science Center. There you'll find biotech research supported by the likes of the Bill and Melinda Gates Foundation improving the nutritional content and drought resistance of African staple crops such as cassava and sweet potatoes. The center is also developing algae and other plants that can produce biofuels, and working on a host of further developments one would expect to excite even the least compassionate among us. But, no, the food police won't have any of it.

In fact, this is all eerily reminiscent of the food progressives' reaction to the green revolution. Nobel Peace Prize–winning scientist Norman Borlaug used research and technology to increase rice and cereal yields in developing countries such as India, Mexico, and Pakistan during the 1950s, '60s, and '70s. The effects were astounding. Wheat yields, for example, jumped more than 250 percent in developing countries from 1950 to 2000. And yet various environmental and international aid groups could not bring themselves to celebrate the fact that millions of people had been freed from starvation— all because the process involved monoculture agriculture, synthetic fertilizers, and technologies that also benefited agribusinesses. Borlaug's response to the criticism is classic:

Some of the environmental lobbyists of the Western nations are the salt of the earth, but many of them are elitists. They've never experienced the physical sensation of hunger. They do their lobbying from comfortable office suites in Washington or Brussels . . . If they lived just one month amid the misery of the developing world, as I have for fifty years, they'd be crying out for tractors and fertilizer and irrigation canals and be outraged that fashionable elitists back home were trying to deny them these things.[14]

Perhaps we Americans can afford to give up a few comforts and pay more for more "naturally" grown food. But this return to nature is pure fantasy. Human interactions with nature have altered animals and vegetation in a way that our ancient ancestors could scarcely have imagined. The food elite's vision of the world would return production agriculture to its state a hundred years ago, to a time when food was far less plentiful and those who did the backbreaking work of farming lived a meager existence. We might like to visit the Amish, but few are clamoring to convert.

THE FOLLIES OF FARM POLICY

Writing a letter to Obama, whom he ironically called "Farmer in Chief," Michael Pollan cried, "Bad news: the food and agriculture policies you've inherited . . . are in shambles, and the need to address the problems they have caused is acute." Pollan proceeded to lay out a buffet of prescriptions to resuscitate the farm sector. Compiling the fashionable demands of the food elite, the Obama administration was, among other things, to support food diversity, use *federal* policies to promote *regional* food economies across the world, utilize agricultural policy to impose environmental standards, revamp school-lunch programs, forgive loans to culinary students in exchange for government support, and renew the 1960s-era policies requiring government purchases of grain.[1]

Reading Pollan's farm policy proposal, one is struck by the vision of an all-encompassing government that knows no bounds so long as its purpose is to provide us fashionable food. Lurking behind the compassion and glowing rhetoric is a deeper reality of food totalitarianism. It isn't bad enough that the government does current farm policy poorly. It must do more, more, more.

Pollan isn't alone in his unease with the present state of farm policy. I share it. We have a federal farm policy that is largely an anachronistic throwback to the past, surviving today mainly through the political power of the farm lobby. Where I part with Pollan and his fellow foodies is in their conclusion that "like so many government programs—what subsidies need is not the ax, but reform that moves them forward."[2] What makes the food police think the government will get it right this time? They like to talk about market failures but are apparently blind to the abundance of government failures. If the process is so corruptible by corporate interests and mega farms, as they claim it is, then Uncle Sam is incapable of working in our food interests, and all the preaching of hope and change is nothing more than smoke and mirrors.

While I am no fan of farm subsidies, I am at least honest enough to tell you that most of what we're told about them cannot withstand scrutiny. Farm policy is like an old leaky faucet. It's annoying and a bit costly but, in the end, relatively benign. By all means, fix the faucet, but don't tell me it's a flood requiring a new en suite! If the food police cannot accurately diagnose the consequences of current policies, we should shudder at their prescriptions for the future. We have a bunch of journalists, chefs, and cookbook authors playing armchair economist without having apparently mastered Econ 101.

One of the constant refrains is that farm policies make

food "too cheap." Pollan tells us that because of fossil fuel use and farm subsidies, "[c]heap food is food dishonestly priced."[3] Modern food production supposedly causes environmental and health externalities and cheapening food serves only to increase the costs borne by third parties. Bittman writes, "Direct subsidies to farmers for crops like corn . . . and soybeans . . . keep the prices of many unhealthful foods and beverages artificially low."[4] The food elite seem to think that all farm policies serve to *lower* the price of corn, soybeans, rice, and wheat. Wrong.

Are farmers such dolts as to advocate policies that primarily *lower* the prices they get paid? Ethanol policies, for example, are hated in many quarters precisely because they *increase* corn prices rather than decrease them—and with dubious environmental benefits. Some of the food police know this. One of their compatriots notes, "Diversion of food to biofuel has contributed to an uncontrolled price rise."[5] Strange, I thought farm policy was keeping corn prices too low, not too high.

What about the roughly thirty million acres of farmland that, through the federal Conservation Reserve Program, farmers are paid *not* to farm? Or the government programs such as food stamps that pay people to eat more food? By far the largest budgetary outlays of the Farm Bill (about seventy cents out every dollar spent) go toward food-assistance programs such as food stamps and school lunches.[6] These programs drive the price of food up, not down. I'm not denying that some farm policies (such as spending on agricultural research) lower food prices, but there is such a dizzying mix of farm programs and supports that for the food police to say conclusively that farm policies unilaterally lower food prices and make them "too cheap" is misleading at best and outright dishonest at worst.

Even the farm commodity programs for corn, soybeans, wheat, and rice that are held in utmost contempt by the food elite have questionable effects on retail food prices. To avoid violating international rules that could provoke retaliatory trade barriers, the U.S. government has increasingly converted farm subsidies to so-called decoupled or direct payments, which are unrelated to the current volume a farmer produces (a trend that might be reversed in future farm bills). With these decoupled payments, the farmers simply get a check from the government that has nothing to do with the amount of corn or wheat or cotton they've planted that year. While these decoupled payments likely grow farmers' bank accounts, they at least let supply and demand function and, as a result, have only small effects on crop prices.[7]

Even the countercyclical payments that go to farmers when prices fall below a price floor have been found to have an effect "on production [that] appears to be negligible."[8] Another recent study determined that "U.S. farm policy, for the most part, has not made food commodities significantly cheaper and has not had a significant effect on caloric consumption."[9] The research shows that completely eliminating U.S. farm commodity programs is likely to have only a small impact on crop prices and production, and could even increase prices for some commodities such as wheat and corn while increasing the production of beef and pork.[10] That's right: getting rid of all current farm programs might *increase* meat production— you know, the sinful process rejected by environmentalists and animal welfare activists alike.

When it all shakes out, what *if* government policies increase the volume of food produced and, on the net, lower food prices? There are benefits of overproduction. After all, it is much riskier to have too little food than too much. According to the UN Food and Agriculture Organization, there are more than a *billion* starving people in the world today.[11] Those concerned about equity should be happy to reduce the price of a necessity for the poor and starving. Moreover, there are positive externalities associated with knowing the cashier, the mechanic, and drivers on the street are well fed and free from worrying about their next meal.

What about us relatively rich folk who can afford to pay more? Even we need to ask what we would get in exchange for the higher prices desired by Pollan and others. A serious defender of the externality argument would have to do the hard work of actually trying to calculate the size of the third-party costs (both positive and negative) rather than run around claiming the sky is falling. In actuality, the food police use the jargon of economics to give an ideological agenda the appearance of real science.

Simply screaming that an externality exists is not proof that something should be done. After all, Nobel Prize–winning economist Ronald Coase taught us that when Farmer Jones's cows trample Sally's field, there is a clear incentive for Sally to negotiate with Jones rather than run to Uncle Sam. Moreover, it is not as though Farmer Jones is an uncaring louse. He is keenly interested in the quality of the land he leaves his children and in their ability to earn a future living. Even beyond his family, Mother Nature has instilled in him and all of us the guilt, compassion, and empathy that cause us to treat our neighbors more benevolently than wielders of textbook definitions of externalities claim.

None of this is to say there aren't some negative environmental consequences that come from agricultural production, organic and conventional alike.[12] But we seem to forget that gasoline is already taxed, as is almost all food eaten away from home; ethanol, biodiesel, and "green energy" are subsidized. Truly earnest advocates of externality-fixing taxes would have to take into account these preexisting food and energy subsidies and taxes, international trade distortions, constantly changing market prices, and all the complex interactions among poverty, health, environment, and food quality to derive an optimal tax/subsidy policy that would likely be obsolete the moment it was computed. Even still, an honest appraisal would have to keep open the possibility that new government action is apt to *cause* an externality rather than solve one.

We are also told that greedy agribusinesses manipulate and capture all the benefits of farm programs. Here's Pollan again: "More even than the farmers who receive the checks (and the political blame for cashing them), these companies [Cargill and ADM] are the true beneficiaries of the 'farm' subsidies that keep the river of cheap corn flowing."[13] It might have done some good actually to look at the economic research studying the true beneficiaries of farm policy. One of the first things students of Intro to Ag Econ are taught is that commodity payments largely flow to the owners of fixed assets, such as landowners, not necessarily to producers. One study estimated, for example, that government payments are responsible for about 45 percent of the total value of Iowa farmland.[14]

So, current landowners definitely benefit. Who else? One prominent agricultural economist estimates that "for every dollar of government spending on farm subsidies, farmers receive about 50 cents, landlords receive about 25 cents, domestic and foreign consumers receive about 20 cents, and 5 cents is wasted."[15] Funny—those evil agribusinesses, the supposed "true beneficiaries," aren't even mentioned. In fact, another study found that farm subsidies have "a negligible influence . . . on the prices of other inputs, such as fertilizers, chemicals, and machinery."[16]

Here we can plainly see the food elite's schizophrenia about farmers. In one breath they speak of the Jeffersonian ideal and the importance of family farms, and romanticize production agriculture. In the next they lament how farmers get massive subsidies from the federal government—in truth only about 40 percent of farms receive government payments—and claim they are either incapable of considering, or unwilling to consider, the health of their land and our children. Indeed, one of Pollan's heroes in *The Omnivore's Dilemma* thinks most farmers are just plain dumb. As he puts it, "Part of the problem is, you've got a lot of D students left on the farm today."[17] The truth is, according to USDA data, "[t]he education level of farm operators is very similar to the educational profile for U.S. householders."[18] Yet we are somehow supposed to believe that the food police residing in residential Berkeley or New York City are more concerned and knowledgeable about the quality of the land in Farwell, Texas, than the flesh-and-blood people who own that land. We must believe that despite large and influential farmer-led commodity groups, farmers are merely subject to the whims of the greedy agribusinesses and don't realize that the policies for which they fight are only handouts for agribusinesses.

The food elite's trick is to divide farms into the good and righteous and the bad and villainous. The food police can maintain their hypocrisy by demonizing "factory" or "corporate" farms. But the reality is that 97 percent of all farms in the United States are family farms.[19] For all practical purposes, there is only one kind of farm in America: the family-owned farm. The food police must be awfully paranoid to worry so much about the few corporate farms—or is it rather a useful scare tactic to throw around the word *corporate* without any context? As if the people who own corporations don't have families, too.

It is true that some family farms are quite big. These large-scale family farms represent only about 7.5 percent of all farms in the United States, but they produce most of the food, accounting for more than 60 percent of agricultural output.[20] But why are they presumed less righteous or wholesome than small farms? If you're concerned about food security, you should be concerned about those who produce the food. A lot of large farmers do well financially, but so do many small farmers. Even small family farmers have average incomes that outpace that of nonfarm households, and they are much more likely to have extra off-farm incomes than their larger family farm counterparts.[21]

The statistic we're frequently shown as demonstrating the injustice of farm programs is that large family farmers receive more subsidies than their smaller brethren. This is of course true in total dollar payouts. What you're almost never told is that as a *proportion* of their agricultural output, small family farms receive more government handouts as compared with large family farms. Small farms get *five times* more government subsidies relative to the value of their output than large farms.[22] One study concluded, "Careful examination reveals that . . . agricultural subsidies appear to be . . . subsidizing small farmers *relatively* more than large farmers."[23] So, at least in terms of agricultural output, small farmers are doing less and the government is giving more. Still, many people think large farms get too much, which is exactly why there have been various forms of farm payment limits since the 1970s.[24]

Again, it must be emphasized that, as a group, farmers large and small do relatively well financially.[25] If equity and fairness are the concerns, we should redistribute money away from farmers of all types, large and small, and toward nonfarmers. Or just get out of the business of trying to pick winners and losers altogether and let the market reward those who produce what consumers want.

Another spurious claim of the food police is that farm policy is to blame for the rise in obesity. To take just one example, the Public Interest Research Group, which was founded by Ralph Nader and once employed Barack Obama, recently released a report concluding that "by fueling the crisis of childhood obesity, the [farm] subsidies damage our country's health and increase the medical costs that will ultimately need to be paid to treat the effects of the obesity epidemic."[26] The media widely interpreted the report as providing a "link" between farm subsidies and obesity—something the food police have claimed for years.

To claim that farm subsidies *cause* obesity, one would have to wonder how it is that Australian obesity rates have risen at roughly the same rate as those in the United States when Australia has no farm subsidies.[27] Indeed, careful statistical studies of changes in farm subsidies over time and across countries show that changes in rates of obesity are unrelated to changes in spending on farm programs.[28] The research is clear: farm subsidies are not the culprit for the rise in obesity.

How can this be when the food police have been telling us the opposite for so long? For starters, as we've already seen, the link between farm policy and food prices isn't as clear as propaganda asserts. It never ceases to amaze me that in the diatribes arguing that farm policy leads to more high-fructose corn syrup (and, voilà, more obesity), the food police never stop to point out, for example, that another government policy, the sugar tariff-rate quota, drives up the price of cane sugar. One estimate suggests that the government sugar programs keep the price of beet and cane sugar 15 percent *higher* than they would be otherwise.[29] A case could be made that current government policies make calories from *all* types of sugar (corn, cane, and beet) more expensive, not less.[30]

You also have to keep in mind that the prices of raw farm products such as corn and wheat make up only a small fraction of the costs in the grocery store. For cereal and bakery products, the farm costs represent only about 7 percent of the total retail price. The other 93 percent of the price of Wheaties is made up of the costs of turning the wheat into flakes, the salary of the truck driver delivering the box, the celebrity-athlete endorsement on the cardboard cover, and on and on.[31] If wheat prices fall 10 percent, the price of Wheaties falls only 0.7 percent. So, *even if* farm policies drive down the price of corn, wheat, and soybeans, this often isn't going to have a big effect on what we pay.

The truth is that in the grand scheme of government largesse, farm policy is just a fruit fly on the massive heap of spending. The USDA spent, in total, about $130 billion in 2010.[32] Over 70 percent of that went to programs such as food stamps and school lunches. Only about 13 percent of the total, or around $16.5 billion, was spent by the agency administering the farm subsidy programs. To put these figures in perspective, note that farm subsidies account for less than 4 percent of the value of all farm output.[33] Also recall that Obama's 2010 federal budget proposal was $3.55 *trillion*. Farm subsidies account for less than half a percent of total federal spending. You can't even make the slice show up on a pie chart. Yes, wasteful and inefficient programs should be cut, but let's not make a Mount Everest out of a molehill.

The fact that the food police can't quite figure out the real effects of farm subsidies hasn't in the least hindered their desire to remake farm policy. The food elite, not understanding the present, believe they have a crystal ball into the future.

At the forefront of the policy agenda of the food elite is a desire not to cut farm subsidies but to expand and reallocate them toward their favored foods. Writing in the *New York Times,* Mark Bittman proposes "support designed to encourage a resurgence of small- and medium-size farms producing not corn syrup and animal-feed but food we can touch, see, buy and eat—like apples and carrots—while diminishing handouts to agribusiness and its political cronies."[34] It is a mystery as to why Bittman supposes the agribusinesses and political cronies supposedly benefiting from subsidies to corn and rice wouldn't also figure out how to profit from federal subsidies to apple and carrot growers. So goes the apparent belief that the favored veggies are morally good and incorruptible while their sleazy wheat and soybean cousins have gone over to the dark side.

If we started subsidizing spinach, artichokes, and arugula, sure, we'd get more of them. But then what? Which agricultural crops do you suppose are the biggest users of water, fertilizers, insecticides, and herbicides? It's not necessarily those evil crops the government currently subsidizes.

If we began subsidizing the crops favored by the food elite, we'd likely see massive increases in chemical use. On average, wheat and soybean farmers use about 0.02 pounds of active insecticide per acre planted; corn farmers use more, at 0.13 pounds per acre. A farmer who plants the same acreage with okra, cabbage, cauliflower, tomatoes, or melons uses about 62 times the pesticide as a wheat or soybean farmer and about 10 times the pesticides as a corn farmer. Lettuce and strawberry farmers use more than 100 times the pesticides as wheat and soybean farmers on a per-acre basis, and don't even mention the citrus growers, who use more than 2,500 times the pesticides!

It isn't just pesticides; it's other chemicals, too. Topping the list of herbicide use on a per-acre basis are crops such as sugarcane, beets, rice, citrus, onions, cranberries, parsley, and asparagus. Cherries, apples, strawberries, tomatoes, and grapes use more than 750 times the amount of fungicides per acre planted as wheat, soybeans, and corn![35]

Chemicals are just one input used to produce food. How

about other resources? Although some vegetables, such as beans, cabbage, and peas, are relatively efficient users of water, it takes more water to produce some types of fruits and vegetables than crops such as soybeans and corn. For example, compared with soybeans, it takes about 4 percent more water to grow tomatoes, 30 percent more water to grow peppers, 39 percent more water to grow sunflowers, 83 percent more water to grow citrus fruits, and 195 percent more water to grow bananas.[36] In their myopic attempt to use government policy to try to get healthy food, the food police seem to forget all the other things that will come with it.

If you buy into Pollan's argument that food is underpriced because of environmental externalities, you ought to think long and hard about Bittman's proposal to subsidize fruits and favored veggies, which are relatively big users of chemicals. Charitably, maybe we should interpret Bittman's suggestion to mean that we should subsidize only organic growers. (But remember, organics use pesticides and fertilizers, too.) The truth is that it is a lot harder to grow and ship crops such as tomatoes and apples than it is to ship corn and soybeans, and even harder to grow them organically. That's why they're more expensive.

I don't know exactly how big the externalities are that concern Pollan, but what is plain for anyone to see is that organic production systems are *much* more expensive than conventional. These are real price differences that reflect real disparities in the resources used in organic food production, not hypothetical claims about the "dishonest" pricing of non-organic. Given the fact that organics often sell at twice the price of conventional, an organic veggie subsidy policy that were to have any substantive impact on what people ate would take a *massive* increase in spending on farm programs to bring the prices down to a level where regular consumers would shift consumption—and all this at a time of ballooning federal government deficits.

The food police claim they are interested in policies that support small farmers but *only if* those farmers produce the kinds of virtuous food accepted by the food elite. They claim they want cheaper veggies, but *only if* they are produced by the kinds of virtuous farmers accepted by the food elite. So we can see that it isn't really about small farmers or health, but about a desire to use farm subsidies to engineer an idyllic food system desired by the food elite.

Here's one other thing you won't hear from Bittman and others who favor subsidies for fruit and veggie farmers. According to a Michigan State University economist, "[i]n 1990, specialty crops growers convinced Congress that they didn't want subsidy payments."[37] That's right, no matter what the food police would have us believe, the large fruit and vegetable producers don't want handouts. Part of the reason is that taking subsidies would invite the government onto the farm to enact restrictive rules of the kind facing corn and soybean growers. But it is not as though these veggie producers are earnest defenders of free markets. Although the matter is currently under negotiation, corn, soybean, and wheat farmers who receive subsidies are restricted from planting fruits and vegetables on the land enrolled for payments. The fruit and veggie growers apparently consider themselves better off allowing government policies to limit competition than explicitly accepting payments themselves. Advocates of the new farm policy would do well to step back and see what happens when we entrust the federal government with the power to tell producers what to grow and what prices to charge. It's a recipe for strategically farming the government programs, not for healthy eating.

One of Pollan's proposals is to establish a strategic grain reserve, whereby the government would "buy and store grain when it is cheap and sell when it is dear, thereby moderating price swings in both directions and discouraging speculation." As if the government's strategic oil reserves have discouraged speculation in the oil market. Rather, as recent history shows with Obama's decision to tap in to the strategic oil reserve, the decision turns from one about market fundamentals to a calculus of politics and reelection strategy. In reality, all that evil speculation is exactly why we don't need a strategic grain reserve.

Don't those big, bad, money-hungry Cargills and ADMs of the world buy up grain now to release later if they expect the price to rise? Or release their current stores of grain now if they think prices will later fall? The natural profit incentives of all the individual actors in the grain market—the farmers, the grain elevators, the traders at the Chicago Board of Trade, and the food processors—cause them *already* to do exactly what Pollan wants to take out of their hands and concede to the all-knowing federal government. Virtually any agricultural economist will tell you that the old government grain storage system never really worked very well, and is even less likely to work well today given the global trade of agricultural commodities.

The optimistic-sounding "strategic grain reserve" is a

guise for what many food police actually want: government-planned supply control. The food elite pine for the old days of the New Deal food policies, when the government bought grain and told farmers what and how much to plant. I don't doubt that with a government security blanket and ample cradling, more small farmers could be kept in business. But at what cost? If the elite's prowess at engineering the agricultural economy had successfully kept the same number of people on the farm today as we had back in the 1930s, what might we have given up? Would Steve Jobs or any number of innovators have been compelled to work with apple trees rather than Apple phones? I'm not saying farming is a bad occupation, but when we distort market incentives, we keep people from knowing where to employ their skills in the places where they can most valuably be put to use.

Farm policy, of course, concerns more than what happens on the farm. Because the bulk of Farm Bill spending happens on food stamps and school lunches, it shouldn't be surprising to see the food police licking their chops to get at these programs. Michelle Obama's signature policy initiative has been the 2010 Healthy, Hunger-Free Kids Act, which, at a cost of $4.5 billion to the federal government, takes more control over children's food at school. From what can be offered in school lunches, to the hallway vending machine, to that old-fashioned bake sale, Uncle Sam now has a major say-so in what our kids eat. Or does he?

Kids don't have to eat what the school offers. As one San Diego school official put it, "Nothing is achieved when money is spent on food that children won't even be able to consume and nothing is more disheartening . . . than to see perfectly good and perfectly untouched food thrown into the trash."[38] As the researchers who have studied the issue could have told Michelle Obama, kids will bring bagged lunches, go off-campus, or just wait till they get home to eat when the lunch lady won't serve them what they want.

Kids and parents are even more likely to shy away from school lunches when administrators have to increase prices to offset the costs required to meet the new law's provision mandating more fresh produce.[39] Higher prices at school mean greater incentives for students to seek lunch elsewhere. They also mean tighter budgets for already cash-strapped schools that are now being forced, by federal law, to trade off double helpings of fruit for extra teachers. The costs of the policy are substantially more than the outlays by the federal government. The much larger costs are the ones borne by all the local school districts.

The truth is that research shows that interventions in school-lunch programs have had very limited effectiveness if any at all.[40] Even when kids are required to eat more veggies at school, they often compensate by eating fewer veggies at home.[41] Studies comparing schools that have removed vending machines to those that have not find virtually no effect on total soda intake and obesity.[42] At least the bill's sponsors can sit back comforted in their do-gooder activities, while janitors in schoolyards across the nation haul out garbage bags of food the elite have mandated but that kids won't eat.

It is not as though public school cafeterias are currently a bastion of free choice. More than 90 percent of public schools participate in the National School Lunch Program (NSLP), whose original goal to increase demand for agricultural

products has been broadened over the years to include nutritional objectives. In fact, now the food police see the NSLP as
an opportunity to sneak in all kinds of new restrictions on
farmers, as witnessed by recent legislative attempts to require
school-lunch purchases to adhere to new animal welfare rules
and local-food sourcing requirements.

NSLP partially funds free lunches and breakfasts for
underprivileged children. But that's not enough to cover the
costs, and local schools typically pick up the remainder of
the tab by eating into their own budgets and by charging the
richer students, who don't qualify. Schools take the funds to
help feed needy kids in their district, but in exchange, they are
required to follow all kinds of rules on what can be served in
lunch lines. Because the federal government reimbursement
rates for NSLP have not kept up with rising food costs, many
schools have started offering separate buffet and snack lines
(exempt from the NSLP rules) that offer items kids will actually eat, to boost the bottom line. So, we have a mess: a convoluted mix of policies that try to get enough calories in the
bellies of poor kids so they're learning during the day and not
starving at night, while simultaneously trying to get the richer
kids, who can have anything they want at home, to eat a few
more carrots at school.

Look, I want kids to eat healthily, too. My father ended his
career as the superintendent of the public school systems in
rural Wellman and Lockney, Texas. On his watch, the Wellman school cafeteria constructed the first-ever salad bar
seen on campus. My point isn't that parents and local school
boards shouldn't think about how to improve children's
health. My point is, *who* is in the best position to make this

determination? It's easy for someone in Washington to enact mandates without any knowledge of the location-specific costs and trade-offs. As a USDA report put it, "Policymakers face hard choices because the children served by NSLP have diverse nutritional needs, making a single policy for all difficult to craft."[43] Will policies requiring local foods, less whole milk, cage-free eggs, or two servings of fresh vegetables cause foods to be more prone to food safety risks or take more time to prepare? Or will they cause kids to go off campus for lunch? Or increase costs and thus class size? These are real trade-offs that only local school districts can adequately address given their own local conditions and knowledge.

Consider, for example, what Christopher Kimball, the founder of America's Test Kitchen and the star of the PBS television show by the same name, had to say about a school district in Vermont: "The school is planting fruit trees. The parents have ponied up their hard-earned money to help purchase local beef instead of what the USDA provides. They are starting to grow many of their own vegetables . . . In true Vermont fashion, these folks are not waiting around until somebody else comes up with a solution. They are just doing it."[44]

These aren't the same choices I would have advocated, but it's not my kids' school and I'm not paying Vermont taxes. So, I say more power to them. The choices made by a school in inner-city Detroit will likely be different, but like the folks in Vermont, the ones in Detroit know more about what their kids want and what they're willing and able to pay to get it.

Then we have the food elite's proposals to reengineer the food stamp program. As if it weren't bad enough that there are people who find themselves in such dire straits that they need extra cash from the government to feed their families, the food elite now wants to use them as guinea pigs in their grand food experiments. In my regular job as a researcher, I can't even ask you a single survey question without having to jump through a bunch of hoops mandated by the federal government to protect the rights of human subjects involved in research, but apparently it is okay with the food elite to play all kinds of games with people so poor they have trouble affording groceries.

Food stamp recipients are already prohibited from using their funds to buy fast food, and now the food police want to keep that money from being spent on items such as soda and other evils; Pollan wants food stamp participants to get a free ride at farmers' markets—he says, "Food-stamp debit cards should double in value whenever swiped at farmers' markets."[45] I want food stamp recipients to eat as healthily as everyone else. But I'm not walking in their shoes. Those of us who have no idea what it's like to have a minimum-wage job and four hungry mouths to feed do not have enough insight into the life of someone in this situation to tell such a person how he should spend his time and money. Rather, it seems the food elite want food stamp recipients to pass a test to prove they know which grocery aisles are sufficiently politically correct enough to enter. I wish we lived in a world where food stamps weren't needed, but reality is harsh. As such, I am not opposed to such meager support for those who find themselves down on their luck, but I can't understand why we'd want to make the people we're trying to help jump through hoops to get it. Food stamps should be an opportunity for compassion, not manipulation.

The food police are right about one thing: it is time for a change in farm policy. But as Hillary Clinton tried to warn us when running for president against Barack Obama, preaching change isn't enough—the question is what *kind* of change we want. Concerns over growing budget deficits will likely mean a different kind of future for farm policy. There are ways to do farm policy more cheaply and effectively. The answer isn't more government control; it's less.

8

THE THIN LOGIC OF FAT TAXES

According to the Centers for Disease Control, we are in the midst of an epidemic: an apparent scourge of fatness.[1] It used to be that health professionals reserved the word *epidemic* for the outbreak of infectious communicable diseases. The fact that obesity is now widely heralded as an epidemic says as much about our pant size as it does the hype surrounding the problem. Fatness is a lot scarier when couched in terms that make you think it's something you can catch from an uncontrollable sneeze on the bus.

Yep, it's true. Americans' waistlines have been growing—for at least a century.[2] Look at the paintings of Peter Paul Rubens circa 1630 and you'll see that being a bit rotund was once socially desirable, signaling, as it did, that you had enough money to eat what you wanted. Today, a few extra pounds aren't seen as a symbol of progress. We're told not only that fatness is unhealthy, but that it is apparently uncouth. Signifying the prevailing mind-set that fatness is failure, the Robert Wood Johnson Foundation released a 2011 report entitled *F as in Fat*. I thought my grades for phys ed ended in elementary school, but whether we want to be judged or not, a report card is coming. Eric Oliver, a political scientist at the University of Chicago, accurately summarized the food elite's attitude: "Running through all these perspectives is a paternalistic condescension toward fatness and fat people—not only do people with this view assume that fatness is inherently bad, but they also presuppose that fat people (that is, minorities and the poor) are too ignorant to know that they should be thin . . . For many people, trumpeting the 'problem of obesity' is an opportunity for them to express both their own moral superiority and their latent class snobbery and racism."[3]

But wait a minute. According to the CDC, obesity is killing us. Back in the late 1990s it was estimated that there were almost 300,000 excess deaths due to overweight and obesity, an estimate that CDC researchers subsequently revised upward to almost 400,000. There was only one problem with the story: it was wrong. When the story broke that the CDC estimates were inflated due to miscalculations, Jay Leno remarked, "Not only are we fat, we can't do math either."[4]

Later, researchers published a study in the *Journal of the American Medical Association* showing that there are, in fact, only about twenty-six thousand excess deaths due to overweight and obesity combined, almost fifteen times fewer than the original estimate. To put the new figure in perspective, note that more than thirty thousand people kill themselves in the United States each year, and there are thirty thousand annual deaths from automobile accidents.[5] Yet I haven't heard much about a national "driving epidemic."

The reality is that pharmaceutical companies and the weight-loss industry make big bucks when fatness goes from a mild nuisance to a certifiable disease. Government agencies such as the CDC and researchers who rely on federal grants also benefit when statistics proclaiming an epidemic can be used to grow research budgets and siphon dollars away from other projects to support the researchers' agendas.

In the mid-1990s, scores of experts were not content with the rise in obesity. They changed the definition to make it look worse than it actually was. Back in 1998, despite the needle on the scale standing perfectly still, twenty-nine million Americans who were previously considered normal weight suddenly found themselves, according to new academic and government wordsmiths, somehow overweight. An epidemic was created with the stroke of a pen.[6]

The data revealed another interesting detail, one often ignored by the anti-fat crusaders. There is evidence that people who are overweight (but not obese) actually live a bit longer than people who are normal weight. Individuals aged 25 to 59 who are slightly overweight (say a 6-foot man weighing 200 pounds) are about 17 percent *less* likely to die in a given year than those who are normal weight (say a 6-foot man weighing 160 pounds).[7] We keep a spare tire on our car for a good reason. As for the one around our waist, apparently Mother Nature put it there for a purpose. So, yes, while being extremely overweight will likely cost you a few years of life, being a bit overweight probably does the opposite. Another shocker: there are more than thirty thousand estimated excess deaths from being underweight. At least in terms of longevity, it's better to be a little plump in your mom jeans than fit in those skinny jeans.

All this squabbling about the correct number of excess deaths is sort of silly. There is no such thing as an "excess death." The researchers have taken all the deaths they can't explain with their models and attribute them to the apparent killer of all killers: obesity. Here's what Eric Oliver had to say about the issue: "The researchers who estimated that obesity costs us 100 million dollars a year did so by calculating all the expenses associated with treating type 2 diabetes, coronary heart disease, hypertension, gallbladder disease, and cancer . . . Once again, they simply assumed that if you got heart disease or breast cancer it was because you were fat."[8]

At an even deeper level, this kind of research simply looks at correlations between weight and death, but we all know that correlation does not causation make. Being too skinny didn't cause some people to die—rather, they were frail because they'd already suffered from some other ailment. What's true for the thin is the same for the fat. The prevalence of the diseases that obesity is supposedly causing has, in most cases, actually gone down over time. One team of researchers found that "[e]xcept for diabetes, [cardiovascular disease] risk factors have declined considerably over the past 40 years in all [body mass index] groups."[9] These are all good signs, except for the case of diabetes.

It is estimated that about 8.3 percent of the U.S. population now has the disease.[10] While we know that the prevalence of diabetes is correlated with body weight, experts do not agree on the cause.[11] It may be that a genetic predisposition toward insulin resistance causes obesity and diabetes (rather than obesity causing diabetes, as is often asserted). That many obese people are not diabetic also casts doubt on the causal link between weight and the disease. Even the American Diabetes Association says, "Type 1 diabetes is caused by genetics and unknown factors that trigger the onset of the disease; type 2 diabetes is caused by genetics and lifestyle factors."[12] It might be more satisfying to point the finger at farm policies and Big Food, but the real culprit might lie much closer to home.

Even if we could figure out what was going on biologically, we do not know with any degree of certainty whether any of the proposed public policies can do anything to reverse the slowly increasing trend. (Diabetes prevalence rates have increased by 0.11 percent annually since 1980.)[13] And while diabetes is a serious problem, there is, as with obesity, a lot of hype surrounding the issue. We hear a lot of concern expressed about the rise of type 2 diabetes among children, but it's actually hard to find firm statistics on the matter. One study in the *Journal of the American Medical Association* said, "The prevalence of type II diabetes mellitus in adolescents is low (<1%)."[14] Information on the CDC website is downright schizophrenic. The CDC first reports that there is no statistical evidence of increasing type 2 diabetes among most children by saying, "A statistically significant increase in the prevalence of type 2 diabetes among children and adolescents was found only for American Indians." But then it immediately reverses course by saying, "Type 2 diabetes in children and adolescents already appears to be a sizable and growing problem."[15] The CDC's own statistics fail to establish evidence of a "sizable and

growing problem," but that hasn't halted the sizable and growing hysteria.

Some health advocates have resorted to downright false-
hoods to promote their cause. A recent ad run by Mayor
Bloomberg's New York City Health Commission featured an
amputee with the caption "Portions have grown. So has Type 2
diabetes, which can lead to amputations." Here's the problem: the man's leg wasn't amputated as a result of diabetes
but rather by Photoshop.[16] As if that weren't bad enough, the
ad ran a week after the CDC reported that the rate of diabetes-
related amputations had fallen by more than half since
the 1990s.[17]

I'm not against our getting to the bottom of the cause of
diabetes or carefully considering strategies to help mitigate
the disease, but we need to look before we leap. And we need
to ensure that the solutions don't cause more damage than the
problem they're designed to fix.

One detail overlooked in these discussions is that the prevalence of many diseases, and even weight itself, is correlated
with age. A nation with a large number of aging baby boomers
is biologically bound to be a nation that, as a whole, weighs
more and is more likely to develop diseases such as cancer and
diabetes. We are living longer, and as a result Americans are
suffering more today from old-age illnesses. In the past, we
didn't live long enough to contract the things that now kill us.

Careful researchers calculate age-adjusted measures of
obesity, cancer, and diabetes incidence (each of which has
risen over the past two decades), but these are statistical ex-
trapolations, not undisputable facts. Ultimately, the question
isn't whether we will die; it's when and from what. Citing

statistics on the number of people dying from this ailment or that—with some supposed link to obesity—sounds scary, but let's face it, we will all die of something someday. But not as soon as we once did.

Since 1990, during a time when obesity was supposedly taking its huge toll, Americans' life expectancies have risen 2.5 years. Even those of us who make it to age 65 can expect to live almost 1.5 years longer than 20 years ago.[18] Not only are we living longer, but the quality of our lives is better.[19] Some of these gains have occurred due to advances in medical technology and increased medical spending. Yet it seems to have escaped the imaginations of many medical health professionals that rising obesity is a rational reaction to better medical technology and to a government-run health care system where the sick no longer have to pay the full costs of treating their ailments.

About 33.8 percent of Americans are today classified as obese.[20] This is up from about 14 percent in the early 1970s. Hence the newspaper headlines such as "Obesity Rates Surge, with No End in Sight."[21] The problem with this rhetoric is that the end *is* in sight and perhaps has already been reached. There has not been a measurable increase in the rate of obesity among men since 2003, and there has not been a measurable increase in the rate of obesity among women since 2000. One group of medical researchers concluded that the prevalence of obesity "may have entered another period of relative stability."[22]

I'm not trying to belittle the problems of obesity or diabetes, many of which are real. I'm asking that we take a step back, eschew all the hype, and think through the consequences of the proposed policies. Clearly there are those who are already overweight and want to shed some pounds. The billions of dollars spent annually on weight loss in this country show that many people wish they weighed less. Some of our desire for weight loss is due to status-seeking, but there are legitimate health concerns, too. We all know and love people who suffer from health problems associated with being overweight. While compassion might be a good reason to do something for a friend or family member, it isn't sufficient justification for government action. When did my weight become someone else's problem?

In a recent nationwide survey, I asked folks whether they thought different groups—such as food manufacturers and restaurants—were primarily to blame, somewhat to blame, or not to blame for the rise in obesity. Only 37 percent of Americans said that food manufacturers were primarily to blame, and only 22 percent ascribed primary blame to restaurants. (Almost as many, at 17 percent, said that the government was primarily to blame.) The big culprits, according to public opinion? Look in the mirror. More than 80 percent of Americans said that individuals themselves were primarily to blame for the rise in obesity, and 60 percent said that parents were primarily to blame.

Despite the overwhelming evidence that Americans see obesity as a personal responsibility, the elites (yet again) think they know better than we do. The Center for Science in the Public Interest has argued, "Obesity is not merely a matter of individual responsibility. Such suggestions are naive and simplistic." CSPI's director of nutrition policy was quoted as saying, "We've got to move beyond personal responsibility."[23]

In regard to sugar intake, one medical university researcher recently asserted, "Everyone talks about personal responsibility, and that won't work here . . . These are things that have to be done at a governmental level, and government has to get off its ass."[24]

So, if we're not responsible for ourselves, who is? And who is responsible for the people who think they're responsible for us? The elite's motive for denying personal responsibility is self-evident. As Thomas Sowell put it, "To believe in personal responsibility would be to destroy the whole special role of the anointed, whose vision casts them in the role of rescuers of people treated unfairly by 'society.' "[25]

The most common retort to the personal responsibility argument is that medical costs, particularly Medicare and Medicaid spending, are spiraling out of control because of obesity. My health care costs rising because of other people's obesity is a reflection more of problems with our health care system than of fatness. It is true that the obese have higher medical costs than normal-weight folk. But the evidence largely suggests that the overweight personally bear the costs of their obesity: they pay higher medical bills, they pay higher insurance rates, and they earn lower wages.[26] That's hardly the sign of hidden externalities. Moreover, if we are going to start counting as a public "cost" the increase in Medicare and Medicaid spending from obesity, we must also count as a public "benefit" the reduction in Social Security, Medicare, and Medicaid spending that comes about from the obese dying sooner than normal-weight folk. If it sounds vulgar to count obesity-related deaths as a public "benefit," one might want to rethink the same logic used to count obesity-related health care spending as a public "cost."

What if we can't reform a health care system that has imprudently made private losses public? If we are stuck with our current health care system, perhaps we should enact government policies to fight the plague of obesity. After all, there are some who believe, perhaps out of a sense of fairness, that "society" should bear the costs of those unlucky souls who lost the genetic lottery to obesity. But *even if* the government is now in the business of deciding optimal waistlines, are the proposals that are on the table even going to work? You might see obesity as a greater public health threat than do I, but surely you wouldn't want to pass costly policies that aren't going to solve the problem.

Let's take a look at the correctives cooked up by the food police.

One of the weapons the fat warriors have been yearning to wield is the ubiquitous fat tax. Writing in the *New York Times*, Mark Bittman calls for the government to "tax things like soda, French fries, doughnuts and hyperprocessed snacks."[27] In his book *The World Is Fat*, Barry Popkin says he wants to "change our eating pattern . . . through taxation."[28] For almost twenty years, Yale professor Kelly Brownell has promoted the so-called Twinkie tax.[29]

Politicians have taken the bait. More than twenty-one states already have taxes that specifically target soda.[30] The city of Chicago recently raised taxes on candies and sodas. Several cities have considered special taxes on fast-food restaurants. And in a sign of what might be coming to the United States, Denmark, despite having one of the lower rates of obesity in Europe, has just implemented a nationwide tax on saturated fat.

With all the hubbub surrounding the various incarnations of the fat tax, one would think there must be a lot of evidence that it would work. Think again. Time after time, economists have shown that fat taxes will have trivial effects on weight. USDA economists estimate that "a small tax on salty snacks would have very small dietary impacts. Even a larger tax would not appreciably affect overall dietary quality of the average consumer."[31] A group of Berkeley economists concluded that "even a 10 percent . . . tax on . . . fat would reduce fat consumption by less than a percentage point."[32] A recent study in *Health Economics* reported that a whopping 20 percent tax on sugar-sweetened beverages would reduce weight by only a paltry two pounds. Reporting in that same journal, I calculate that people would have to be willing to pay an astounding $1,500 per pound of weight lost for this kind of tax to make sense.[33] Another study analyzed obesity rates across locations with varied soda tax policies and found "existing taxes on soda . . . do not substantially affect overall levels of soda consumption or obesity rates."[34]

Fat taxes don't work because food purchases are rather insensitive to price changes. We spend a relatively small portion of our income on food, and this means that food price changes, at least at current levels, don't dent our budget enough to induce a change. That is, of course, unless you're poor. Which is another reason fat taxes are a terrible idea. They are regressive—meaning the costs are disproportionately borne by the most disadvantaged. Food is a necessity, and it comprises a larger share of the poor's budget. As a result, fat taxes penalize those who can least afford it.

Fat taxes are also impotent because, as consumers, we can shift consumption to untaxed ingredients that may not be as healthy. Taxes on sugared sodas are likely to increase consumption of alcoholic beverages and naturally sugar-filled fruit juices. Taxes on saturated fat will reduce intake of healthy items such as avocados, salmon, and cashews. And as I've shown in research published in the *Journal of Health Economics,* taxes on fast-food joints and restaurants may actually cause us to *gain* weight by increasing the consumption of fattening foods at home.[35]

If we really wanted to curb fat through taxes, many economists would tell you, it would probably be more effective to tax fat *people* than fat *food.* Most fat-tax advocates are rightfully appalled by such a proposal, but I have yet to see a compelling reason why taxing fatty foods is any more righteous than taxing fatty folk. If obesity imposes an externality on others—as fat-tax advocates claim it does—then the only "fair" thing to do is to tax the source of the externality. The food police don't want to tax fat people, because it shifts responsibility from the "toxic food environment" to the individual. It's better to have an ominous-sounding enemy such as Big Food to blame than to look as if you're discriminating against overweight people. But food prices are part of our environment. And so, too, is the price for being fat. Making someone who wants to eat Doritos pay 20 percent more (or switch to eating something that wasn't chosen before the tax) is a real cost on the real people the fat taxers say are suffering. Fat-tax advocates would rather promote a less-efficient policy than fess up to the fact that they're imposing big costs on the same people they claim to be helping. For the record, I'm against fat taxes—and that goes for the food *and* people varieties.

Despite all the evidence against fat taxes, the food police are undeterred. Why? Because they know that taxes bring in revenue to grow the size of the government. As the Berkeley economists put it, "A fat tax is an effective means to raise revenue."[36] If anyone's growing waistline should be carefully scrutinized, it is Uncle Sam's. Yet the food police see fat taxes as an opportunity to fund their preferred plans. Bittman proposed that "[t]he resulting income should be earmarked for a program that encourages a sound diet for Americans by making healthy food more affordable and widely available."[37] Others say the tax proceeds should go toward government information campaigns. For example, one New York assemblyman pushing a statewide fat tax said it should go forward "as long as the revenue is directly targeted and used to address a healthy lifestyle, and not to fill a budget gap."[38]

The discussion reminds me of Al Gore's Social Security lockbox. The trouble is that our Social Security funds are *not* locked in, and today's taxpayers are funding yesterday's workers. If we can't earmark something as important as Social Security receipts, what makes healthy-eating advocates think we could do the same for fat-tax revenues? Senators, defense contractors, and the underwater Post Office would be among the beneficiaries of extra government revenue that was dreamily presumed to fight Big Food. Remember, too, that only about 2 percent of the funds from the Tobacco Settlement, which were promised to fight smoking, actually went for that purpose. The states spent the rest to fix budget shortfalls.[39]

Never mind whether we can lockbox fat-tax revenues, and set aside the issue of whether it is even a good idea to grow government revenues at all, where is the evidence that a government "program that encourages a sound diet" would even work? The elite's favored information and education campaigns have been remarkably ineffective. If the billion-dollar weight-loss industry cannot, with all its marketing expertise, persuade people to lose more weight, what makes the government think it can? There are already many public-sector information initiatives against obesity, not to mention Michelle Obama's Let's Move! campaign, and it is unclear what effect yet another campaign would have.

Existing government information policies have had only minimal effects—as if most of us didn't already know that carrots were healthy and pizzas weren't before Michelle Obama put healthy eating center stage. The USDA's food pyramid, once familiar to all schoolchildren, did not stem the rising tide of obesity (and according to low-carb advocates, it actually contributed to it), and it is unclear that changing the pyramid to a plate, in its latest incarnation, will have any substantive effect on what we eat. Mandated nutritional labeling on food items in grocery stores has had very little, and in some cases a deleterious, effect on the healthfulness of food choices.[40]

The new mandatory calorie labeling law for restaurants buried in Obama's health care reform bill has been shown to have little effect on what people order, and sometimes leads to counterproductive choices. My own research shows that the labeling's biggest effect may be to reduce the profitability of restaurant owners.[41] As one group of researchers indicated in the pages of the *New York Times,* "To pin our hopes on calorie posting is bad lawmaking based on poor reading of science."[42] One Harvard economist called calorie labeling a "revenue-less tax." A Berkeley economist concluded, "The causal-chain from mandatory information consumption to improved health outcomes is so weak that one wonders whether it is worth making people feel bad about themselves."[43]

Faced with the reality that fat taxes and information campaigns are unlikely to work the way they're advertised, the food police have resorted to attacking one of our most cherished liberties: the freedom of speech. Here is Popkin in *The World Is Fat*: "If we still had the Fairness Doctrine, and if public health advocates successfully applied this to the billions of dollars that the food and beverage industries spend on advertising, a great deal of money would be available for counter-advertising, research, and other activities that promote healthier diets. But it's twenty years since the repeal of the Fairness Doctrine."[44]

Yep, even the advocates of fat taxes know that such taxes won't meaningfully reduce weight, but they pine for them anyway, so they can use the funds to create their own commercials. What is "fair" about a government that can forcefully take our money through taxation and then turn around and give it to some elite food preachers, all the while prohibiting food industries from spending money they've *earned*? The original Fairness Doctrine could at least claim some aura of legitimacy at a time when most broadcasting took place over public airwaves, but that logic is now gone. Moreover, when the FTC repealed the Fairness Doctrine in the late 1980s, it did so because of evidence that the law quieted dissension and controversial speech.

The food police argue that there is a dire need to counter-act the advertising of Big Food. But companies selling high-fat, high-sugar foods are already living in a media environment that undermines their goal to sell more. What adorns the covers of the magazines lining the checkout stands at the grocery store? Overly thin celebrities and supermodel waifs. Turn on the TV and who appears? Skinny news anchors and commercials featuring people with six-pack abs. The prevailing cultural and advertising message is clear: thin is cool, fat is not. So it's apparently okay to make my wife feel guilty about her weight when she checks out at the supermarket but not okay for Kelloggs to encourage her to eat frosted shredded wheat?

One has to wonder why a company as manipulative and powerful as McDonald's would introduce such failures as the McLean Deluxe. The food nutritionist extraordinaire Marion Nestle is disturbed that Big Food is lurking in labs trying to conjure up recipes that taste good and then, shockingly, advertise to get us to try them. If McDonald's isn't in the business of trying to sell me things that taste good again and again at a price I'm willing to pay, I'm not sure why the company exists. What bothers me is that Nestle is so troubled that there is a company that serves this purpose.

The food police seem to forget that Big Food competes not just on taste but on health, too. It was actually the *removal* of government bans on advertising health claims that led to product innovation and healthier offerings.[45] If consumers care about health, then food manufacturers have an incentive to deliver what we want. Look around the next time you're in the grocery store and you'll see the shelves overflowing with low-sodium, low-carb, high-fiber products. McDonald's,

Wendy's, Burger King, and the whole lot offer salads, grilled chicken sandwiches, and fruit to kids. So, yes, of course Big Food tries to get us to buy more of what they make, as does any other company. But as long as you care about healthfulness, so will they.

The proposals to restrict speech have gained significant traction in the Obama administration. A partnership of four federal agencies has formed something called the Interagency Working Group. The group has issued guidelines that would severely limit marketing of certain foods, especially ads aimed at children. The guidelines are so broad as to prohibit your local doughnut store from sponsoring Little League T-shirts. Two U.S. congressmen appalled at the group's guidelines argued, "American favorites such as Honey Nut Cheerios, Cheez-It, Barnum's Animal Crackers, peanut-butter-and-jelly sandwiches and even celery and bottled water couldn't be marketed or advertised to children in any manner."[46] I don't want my kids to be unhealthy, but I have a remote control and I know how to use it.

Kicking Tony the Tiger off *Dora the Explorer* may not seem like such a big deal, but if you want your kids to have cartoons to watch, somebody has to pay for them. If Kraft and General Mills are no longer allowed to fund commercials for children's shows, that means the shows will be less profitable and therefore more infrequent. You might think that's a good thing—unless you work at Disney or Nickelodeon or even NBC. More important, a government that can tell companies what they may and may not say with their own money is treading on very dangerous territory. As an editorial in the *Tampa Tribune* noted, "They get rid of Snap, Crackle and Pop, would they next target the AFLAC duck and the GEICO gecko?"[47] These kinds of rules promote the worst kind of crony capitalism, because they force businesses to spend more time attending to the bureaucrats than to their customers.

Case in point: Michelle Obama recently appeared on the White House Lawn with the CEO of Darden Restaurants (owner of the Olive Garden and Red Lobster), whom she praised for promising to reduce calories and sodium by 20 percent over the next ten years. Why would a company such as Darden do something like this? Why wouldn't it? When the Obama administration can selectively hand out waivers for their health care mandates and arbitrarily pick recipients of bailouts and green job grants, it's pretty clear where the incentives lie. The do-gooders got to be seen doing good, while Darden got free publicity for something it could have done of its own accord. A 20 percent calorie reduction can be achieved with a 20 percent reduction in portion size, so Darden can cut costs and avoid looking cheap by appearing healthy. In the end, it's all about the photo op. Besides, who ten years from now is going to check whether the company lived up to its promise?

The food police not only want to restrict food company speech, they also want the government to wage a propaganda war on its own citizens. Time and time again the food police cry for a antitobacco-like campaign against Big Food. Kelly Brownell has taken to equating Ronald McDonald with Joe Camel, and recommends that food activists "develop a militant attitude about the toxic food environment, like we have about tobacco."[48] Marion Nestle concurs with the analogy when writing, "Like cigarette companies, food companies . . . expand sales by marketing directly to children."[49]

There's a big problem with this analogy: you *don't* have to smoke but you *do* have to eat. A smoker can avoid cigarette taxes by becoming a nonsmoker, but a non-eater is a dead eater. Most nutritionists would admit that eating a few Twinkies or Doritos in the context of an overall healthy diet won't do much harm, but who wouldn't say having a few smokes a week is a bad idea? Thus has begun the language war. "Big Food" is meant to conjure Big Tobacco. Sodas are called liquid candy. Bittman repeatedly refers to "real food" as though some foods were fake—as if béarnaise sauce made in three-star restaurants isn't also "hyperprocessed" food. And now there is a full-on attempt to prove that sugar and fat are addictive. (Somehow these so-called food experts are surprised to learn that the reward centers of our brain are activated when we eat something tasty.)

Given all this, it is instructive to recall that "[i]n a June 1994 newspaper ad that criticized proposals to sharply raise tobacco taxes, R.J. Reynolds said, 'Today it's cigarettes. Will high-fat foods be next?' Antismoking activists traditionally responded to this sort of slippery-slope argument by insisting that cigarettes were unique, 'the only legal product that when used as intended causes death.' To suggest that anti-smoking measures might pave the way for attacks on cheeseburgers and ice cream, they said, was just silly."[50] Looks like silly has officially arrived.

Without solid footing backing their anti-fat proposals, the food police have a powerful rhetorical strategy. Fat taxes won't work? Neither will educational campaigns? But it's all for the kids! As if emotional appeal were a good substitute for some kind of rationally compelling argument. Clearly, the presumption is that parental responsibility is inept. Yet some-how I am supposed to believe that a nutritionist I've never met cares more about my children than I do?

If I can't say no to Toucan Sam, I'm likely to have big-ger problems on my hands than my kid's pant size. However alarmed we might be about childhood obesity, it is prudent to think long and hard about where the arguments abrogat-ing parental responsibilities are heading. Even an outlet as respectable as the *Journal of the American Medical Association* recently published a commentary by an M.D./lawyer team ar-guing that obese children should be taken from their parents and placed in foster care.[51] Apparently obesity is now seen as tantamount to child abuse. If you're willing to give up respon-sibility for your children, the food police are willing to take them—by force. As is so often the case with the food police, they are willing to take such a radical move without any evi-dence that the children, having been seized from their homes and handed over to the care of the state, will go on to lead overall better lives. Your broken home is the price the food police are willing to pay so Junior can lose a few pounds.

In their war on fat, the food police rarely stop to consider carefully why we have gotten fatter in the first place. They blame Big Food and agribusiness farm policy conspiracies, but the research shows that none of this stuff has really had much impact on obesity. In fact, one has to wonder why obesity is rising almost everywhere in the world if it is *our* farm poli-cies and *our* evil food processors causing it all. Even a country such as France, which the food elite so wish to emulate, has seen obesity rates more than double in recent decades.[52] If one wants to understand why weight is rising, you have to look for trends that transcend cultures and national boundaries.

One of the key contributors to obesity is exactly that program the food police wish to reenact: the war on tobacco. Smoking tends to reduce weight, and in an era when smoking has dramatically declined, we should only have expected rising weight. One study analyzing a wide range of factors such as changes in food prices, number of restaurants, urbanization, and type of employment concluded, "We find that cigarette smoking has the largest effect: the decline in cigarette smoking explains about 2% of the increase in the weight measures. The other significant factors explain less."[53] If we have to choose between weighing a little more and smoking a little less, the healthier choice is clear. Yet the food police deny these trade-offs exist or are even relevant.

There is something even more important than tobacco lurking behind our weight gain. It's technology. According to a group of Harvard economists,

> there has been a revolution in the mass preparation of food . . . Technological innovations—including vacuum packing, improved preservatives, deep freezing, artificial flavors and microwaves—have enabled food manufacturers to cook food centrally and ship it to consumers for rapid consumption. In 1965, a married woman who didn't work spent over two hours per day cooking and cleaning up from meals. In 1995, the same tasks take less than half the time. The switch from individual to mass

preparation lowered the time price of food consumption and led to increased quantity and variety of foods consumed.[54]

Don't let the nutritionists fool you. There is no such thing as a single ideal weight. There is more to life than what appears on our bathroom scale. It is only reasonable that we eat a little more food when it costs less and give up manual labor when an air-conditioned office job comes along. We can't disentangle all the bad stuff we don't like about obesity with all the good things we enjoy, such as driving, eating snacks, cooking more quickly, and having less strenuous jobs. Yes, we can have less obesity, but at the cost of things we enjoy.

When you hear we need a fundamental change to get our waistlines back down to where they were three decades ago, beware that it might take a world that looks like it did three decades ago. I for one am not willing to give up power steering, microwaves, and inexpensive takeout food even if my pants now fit a little more snugly. So if someone asked me what I was doing to fight the war on obesity, my answer would be the same as that of University of Chicago law professor Richard Epstein: "I play basketball."[55]

THE LOCAVORE'S DILEMMA

In the summer of 2008, Barack Obama traveled to Denver amid much hype and fanfare to accept the Democratic nomination for president. While most will probably remember his soaring rhetoric set against a backdrop of adoring fans packed into a football stadium, fewer will recall that Obama's belly was probably full of local food. Food local to Denver, that is. Denver welcomed fifty thousand nonnative people to Colorado for the Democratic National Convention, but nonnative *food* wasn't invited. The greening director of the DNC required caterers to ensure that 70 percent of their food had been grown either locally or organically. He was apparently sufficiently empowered not only to bar Nebraskan and Californian delicacies but fried foods, too. Fortunately there were plenty of good restaurants outside the convention center where people could get a tasty meal.

The consequences of the local-foods policy were as predictable as the media's reception of Sarah Palin. One local caterer confessed, "We all want to source locally, but we're in Colorado. The growing season is short. It's dry here. And I question the feasibility of that." With so many contractors chasing so few local and organic foods, the result was inevitable: higher food prices for Coloradans and lower profits for local caterers. Even Denver's greening director recognized the problem: "The organic and local route may be more costly, but the committee thinks caterers will find ways to comply and still make a profit ... It takes some creativity because some of these things are more expensive."[1] With lower profits for suppliers, higher prices for consumers, and questionable environmental and health benefits, the DNC inadvertently offered up exhibit number one in the case against the food elite's trendiest policy.

Once the food elite learned they couldn't keep agribusinesses out of organics, a new system advancing the anticapitalist food agenda had to be found. In 2007, *Time* magazine printed the answer right on its cover: "Forget Organic; Eat Local."[2] It is the purported answer for every fashionable food problem. After *The Omnivore's Dilemma* romanticized the alternative food system, Pollan wasn't content to let us consider his arguments on their own merits. Local foods had to be thrust upon us by federal policy. He later wrote, "In the same way that federal procurement is often used to advance important social goals (like promoting minority-owned businesses), we should require that some minimum percentage of government food purchases—whether for school-lunch programs, military bases or federal prisons—go to producers located within 100 miles of institutions buying the food. We should create incentives for hospitals and universities receiving federal funds to buy fresh local produce."[3]

As if we were still driving carts and buggies and our kids weren't competing in a global marketplace against the tech-savvy Japanese, Kim Severson, former food writer for the *New York Times,* imagines rewriting school curricula to require that "students learned math by laying out garden plots and biology . . . taught as students prepared soil, gauged the weather and harvested food . . . Recipe writing might become an English lesson. Digging weeds and building fences would count as physical education."[4]

Local foods aren't just promoted as good economic policy; they represent a worldview. Alice Waters believes "that the most political act we can commit is to eat delicious food" produced in a way consonant with her vision of the world.[5] James McWilliams, author of *Just Food,* reveals that "buying local is a political act with ideological implications."[6]

The politicians are on board. After U.S. Democratic representative Chellie Pingree introduced the failed Eat Local Foods Act, which would have subsidized local foods in school lunches, she teamed up with fellow Democratic senator Sherrod Brown in an attempt to include the more ambitious Local Farms, Food and Jobs Act in the 2012 Farm Bill. The Obama administration has signed into law a rule giving preferential treatment to local suppliers in contracts for school-lunch purchases. Forty-five state governments fund farm-to-school programs, encouraging cafeterias to buy from local sources. The USDA has unveiled the Know Your Farmer, Know Your Food initiative, offering grants, loans, and other support for local foods, and spends $10 million annually on the Farmers

Market Promotion Program. And of course, Michelle Obama spent our tax dollars digging up her back lawn to feed the White House.

Too bad nobody's stopped to ask what business Uncle Sam has telling us where to buy our food. Don't get me wrong. Fresh fruits and veggies are tasty in the summer. When the season rolls around, I'm a regular at my local farmers' market. As a kid, I spent many a summer day in kind neighbors' fields picking surplus peas, corn, and squash. But somehow the personal pleasures that come with the harvest season have been twisted into an argument that I have to pay for others' local-food purchases. Maybe you think it would be a good idea to persuade folks to eat more veggies. Fine. But why *local* veggies? The locavore's answers hold about as much water as a colander.

In 300 AD, Christian leaders met in Nicaea to develop a creed, a list of tenets one must believe in order to be fully considered a member of the faith. The Church of Locavore likewise requires adherence to a set of core doctrinal beliefs. Worthy converts who have attained the title of locavore must place their faith in the Church's doctrines, and with continual chants and incantations, the worthy can hope to accept them as truth. Fortunately, the Age of Enlightenment confronted the Christian Church and brought about less dogmatism and more tolerance. By confronting the veracity of their tenets, I hope to have the same effect on locavorism.

THE FIRST TENET OF LOCAVORISM: LOCAL FOODS ARE GOOD FOR THE ECONOMY

Michael Pollan recently told Bill Moyers, "Let's require that a certain percentage of that school lunch fund in every school district has to be spent within 100 miles to revive local agriculture, to create more jobs on farms, to, you know, rural redevelopment. . . . You will have, you know, the shot in the arm to local economies through helping local agriculture."[7]

I don't doubt such a policy can enrich *your* local farmers—at least temporarily. But at a cost to whom? Let's say the city of Northville requires that its schools and hospitals buy only local food. That's good for the farmers of Northville. Farmers of Southville are hurt, however, when they can no longer sell to Northville. And when Southville foodies get around to enacting their own local-food laws, Northville farmers now lose a market, too. Locavores only imagine the first step and never stop to consider what happens when everyone is pursuing their own inward-looking food policies.

In *Grand Pursuit,* Sylvia Nasar summarized the views of great economists from Keynes to Hayek on international trade when writing, "Erecting trade barriers, and clamping controls on capital outflows might be effective for reducing balance-of-payment deficits . . . and pumping up government revenues. But if everybody adopted the same tactics, the eventual result would be universal impoverishment and unemployment."[8] Economists of all stripes have known for decades that such isolationist policies are terrible ideas for governments trading across imaginary lines defining countries, but there is no fundamental difference when it comes to policies that try to restrict trading across the imaginary lines that define local towns, cities, and states.

A policy that aims to enrich local farmers (even if temporarily successful) keeps some people on the farm who would have found their skills more usefully employed elsewhere.

This might be good local development policy for Northville, but national economic policy should do anything but direct people toward industries where less labor is now needed.[9] Unless the aim is to go back to a system where feudal lords kept their serfs confined to fiefdoms, entrepreneurial farmers will move themselves and their skills (rather than their food) to the locations where they can earn the biggest buck. It is naive to assume that current local farmers are the same people who will ultimately benefit from local-food policies, and it is wishful thinking to presume that successful local farmers won't consolidate and adopt cost-saving technologies to drive the inefficient out of business.

Local-food policies also completely ignore the undeniable fact that Northville is relatively better at growing some things than Southville, and vice versa. Forcing Northville agencies to buy only food grown in Northville means making them pay more for certain types of food than they otherwise would— and when Southville enacts the same policy, it keeps Northville farmers from being able to take advantage of their ability to grow some foods more efficiently.

This is the stuff of Econ 101, and no matter where the subject is taught, whether it be Harvard or South Plains Junior College, a student is almost certain to learn about the wonders of specialization and comparative advantage. Economists may disagree on whether stimulus spending is needed to jumpstart an ailing economy, but outside of a few cranks, you won't find many denying the advantages of free trade—that by letting people specialize in making those things they are relatively good at making and then trading with others, we're all richer than we would be in a world where Northville consumers buy only from Northville producers.

Another lesson of Econ 101: economies of scale. Because of the need to buy tractors and other large pieces of equipment, the larger a farm grows, the more efficient it tends to become. The per-unit costs are often much lower on larger farms. One study of Illinois farms showed, for example, that average total costs were 82 percent lower on soybean farms and 38 percent lower on corn farms that were larger than nine hundred acres as compared with those that were smaller than three hundred acres.[10] Another study showed that average incremental costs were 85 percent lower on dairies with herd sizes greater than 2,000 head as compared with dairies with fewer than 30 cows.[11] What does this have to do with local foods? In order for farms to take advantage of these economies of scale, they need a large market. Policies that restrict sales to certain locations are policies that limit the size of the market. And policies that aim to support smallness per se are policies that tend to promote inefficiency.

Pollan probably knows that local foods cost more, which is why he encourages us to "pay more, eat less."[12] What he doesn't admit is that local foods often cost more because their production is less efficient. Less-efficient production obviously doesn't bode well for the environment; it means we're using up more inputs to create the same amount of output. But it also means that, as consumers, we are poorer than we would otherwise be. So, *even if* local-foods policies enrich local-foods producers, they impoverish local consumers.

Sure, many people are willing to pay more for local foods,

but if their willingness to pay extra was greater than the cost difference, there would be no need for all the cajoling and restrictions on where schools and other institutions can source their foods. By forcing policies that consumers aren't fully willing to pay for, the locavores are doing the equivalent of burning dollar bills. That's hardly a prescription for economic development.

Yet the locavores insist that local foods are somehow more efficiently produced.[13] In *The Omnivore's Dilemma*, Pollan tries to illustrate the productivity of a small, "sustainable" farmer who works night and day to avoid buying off-farm inputs and sells locally. In what can only be seen as an attempt to overwhelm the reader with the "impressively efficient" farm, Pollan tells us that, in a year, the Virginia-based Polyface farm produces 30,000 dozen eggs, 12,000 broilers, 800 stewing hens, 25,000 pounds of beef, 50,000 pounds of pork, 800 turkeys, and 500 rabbits.[14] Add it all up and it amounts to about 300,000 calories per acre produced. Impressive! Or is it?

When an efficient large-scale farmer plants his field with corn, he yields over 15 million calories per acre. That's over 5,000 percent more calories per acre than the Polyface farm! And with a lot less labor. In fact, the locavores' arguments about efficiency tend to treat the most precious of all resources, our time and labor, as if they were worthless and had no value outside the farm. If you like the guys at Polyface and want to visit them by exporting yourself to the food on their farm (even though they won't, ironically, do the reverse), I'm all for it. But don't tell me their system is more efficient or will somehow enrich our nation if mimicked everywhere.

THE SECOND TENET OF LOCAVORISM: LOCAL FOODS ARE GOOD FOR THE ENVIRONMENT

That local foods travel fewer miles is taken as self-evident proof that they are thus better for the environment. Self-evident, that is, until one does even the least bit of critical thinking. The truth is that the number of miles a food travels has very little to do with its overall environmental impact. One extensive review in the journal *Trends in Food Science and Technology* concluded, "It is currently impossible to state categorically whether or not local food systems emit fewer [greenhouse gases] than non-local food systems."[15]

Just to take one example, researchers have found that Londoners would use less energy and emit fewer emissions if they bought lamb raised half a world away, in New Zealand, rather than locally.[16] How can it be four times more efficient to buy lamb that's traveled ten thousand more miles? Comparative advantage and the low costs of sea transportation. New Zealand is naturally endowed with abundant, grassy land and a climate that is especially suitable for rearing sheep. Better New Zealand weather means more grass and less need for supplemental feed and fertilizer than are required to produce British sheep. When you add up the energy savings from the more efficient New Zealand lamb production, it more than offsets the extra energy required to get the stuff to London.

What's true of sheep is also true of all kinds of other commodities. Tomatoes, for example, require ample amounts of warm weather to reach full potential. Outside a few months in the summer, the research shows that it is more energy efficient for northern European cities to ship tomatoes up from the south than to grow them locally in greenhouses.[17] Natural differences in weather, availability of tillable land, and other resources convey comparative advantages to specific regions that add up to significant differences in the amount of energy required to produce certain food. There are good economic reasons why corn tends to be grown in Iowa, wheat in Kansas, potatoes in Idaho, peanuts in Georgia, and grapes in California. It is because these crops can be grown there at higher qualities and in greater quantities using fewer resources than in other locations.

Moreover, successful farmers of a crop that benefits from a region's natural resources can often expand and gain economies of scale. The larger size provided by scale economies results in more efficiency and less energy use. One environmental economist suggests the possibility that "larger producers, because they are more energy-efficient, can transport food longer distances and still have a smaller environmental impact than smaller, local producers."[18]

Of all the global warming impacts that are said to come from food consumption, only 10 percent is due to transportation, whereas 80 percent is a result of activities on the farm.[19] The implication for those worried about global warming is clear: to reduce the carbon impacts from food consumption, one should grow food on farms where production is most efficient and then ship it to the consumer.

While locavores obsess about the distance foods travel to get to them, seldom do they emphasize how far *they* have to travel to get to the food. Shipping food by boat or barge is incredibly efficient, and while shipping by semi is worse, it isn't all that bad. The really inefficient part of food transportation comes from our cars' emissions from the home to market. One study estimated "that if a customer drives a round-trip distance of more than [4.15 miles] in order to purchase their organic vegetables, their carbon emissions are likely to be greater than the emissions from the system of . . . large-scale vegetable box suppliers. Consequently some of the ideas behind localism in the food sector may need to be revisited."[20] So, unless locavores are one-stop-shopping at the farmers' market, chances are that the extra stop to buy local foods is creating more emissions than any that might have been saved getting that food to market. That's also the reason Harvard economist Ed Glaser says that urban agriculture is environmentally harmful: by diverting living space to crops, it increases commuting distances, which causes much more environmental harm than any benefit accruing from the local fare.[21]

As far as the environment goes, it is important also to recognize that "buy local" is a cause, not a certifiable production practice. *Some* local producers use organic and low-tillage production methods, but many do not. At least with organic, there is a certifying body that requires adherence to certain standards to attain the organic label. There is no standardization with local. Some locavores think that's a good thing. But one consequence is that you can never really be sure (even if you ask) that a local tomato was grown with more or fewer pesticides or in a way that caused more or less soil erosion than the tomato traveling cross-country. In fact, one of the main goals of locavorism is to increase consumption of vegetables, and as I have already revealed when discussing farm policy, growing fruits and veggies can require more pesticides and fungicides than growing grains. More chemical usage hardly implies improved environmental impacts. Maybe you think that means you should buy organic, but what does that have to do with local?

And what happens to all the food a local farmer can't sell? One study showed that a large percentage of the crops brought to farmers' markets were never sold and thus simply perished. More than half the tomatoes suitable for market had to be thrown away because nobody bought them. The local farmers also had trouble growing produce of an acceptable size and appearance to consumers, and thus another 20 to 30 percent was thrown away even before market. This kind of waste is not nearly as problematic on large-scale commercial farms. As the authors put it, "[A] large proportion of each type of squash (especially the zucchini squash), cucumber, bell pepper, and okra could not be marketed due to poor quality or because it was too large for the market. In a large-scale production state such as California, much of the defected produce could be salvaged by frozen-food processors or possibly by a food cannery."[22]

Because they share weather and temperature, all farms within a given region are likely to have their produce come to market around the same time. In a world with regional and international trade, that isn't a big deal, as the surplus can be shipped out to other locations. But in the locavore's world, the result is inevitable: spoilage and waste—not something typically prized by recycling environmentalists.

THE THIRD TENET OF LOCAVORISM: LOCAL-FOOD SYSTEMS PRODUCE A MORE ROBUST, SECURE FOOD SUPPLY

In a diatribe against food imports (although I presume he's excepting Italian olive oil, Belgian chocolate, and French Champagne), Mark Bittman writes, "I'm not a jingoist, but I'd prefer that more of my food came from America. It'd be even better, really, if most of it came from within a few hundred miles of where we live. We'd be more secure and better served, and our land would be better used."[23] The idea that a more local food supply is a more secure food supply rests on shaky premises. As we've already seen, local food production will often mean less-efficient food production, which in turn means less food. That's fine if you're trying to cut calories, but food security means that food is available when you want it.

It would be foolish to invest all your retirement savings in a single stock. The financial experts tell us to diversify. And if we shouldn't keep all our financial eggs in one basket, the same goes for hen-laid varieties. One of the things that make farming unique compared to other businesses is its unusually large reliance on the weather. An unexpected drought, rain at the wrong time, an early freeze, or a hailstorm can devastate a whole farming community or even an entire region. While farmers protect themselves financially against these kinds of risks by buying crop insurance, what about the food consumer?

In a world of extensive food trade, there is little need to worry about the consumers of Northville if their farmers face a flood, because Northville consumers can readily buy their food elsewhere. But a world where the locavores have tied the hands of farmers in Northville, Southville, Eastville, and Westville to supply only their local consumers is one where a weather disaster in one location can have dire effects on the consumers who live there. After all, agriculture is a business that requires long production lags. A farmer can't produce po-tatoes on a whim; it takes months of planning and foresight.

The world sought by the locavores isn't more food secure, it's riskier—in terms both of the availability of food and of the prices we'd have to pay.

The locavores want us to imagine that only those small farmers planting heirloom tomatoes are adding to the diversity of regional cropping systems. The reality is that the so-called mono-cropping systems that are the bane of the locavore's existence are far less "mono" than they let on. A large farmer who plants corn this year typically rotates to soybeans the next. Even continuous corn or soybean farmers will often plant a fall cover crop of wheat, rye, or barley. And as I've already shared, many of the big farmers do use the low- or no-tillage production practices praised by the locavores. It is true that the crops these farms plant are not as diversified as those on the farms prized by the locavores, but there is much more temporal and geographic diversity in the global food system than locavores often admit.

THE FOURTH TENET OF LOCAVORISM:
LOCAL FOODS ARE HEALTHIER

Most of us could stand to eat a few more fruits and veggies. But what does that have to do with local food? The conclusion that local food is healthier is a non sequitur. It depends on what local food you eat, when you eat it, and what you do with the extra time and money you don't spend seeking out local products. In short, eating local has nothing to do with eating healthily.

It is true that many fruits and vegetables are most nutrient-rich immediately after being picked ripe. But large growers and processors often quickly freeze just-picked produce to preserve freshness. In fact, frozen vegetables are typically more nutrient-dense than fresh veggies that have been off the vine for several days. One of the big problems for locavores participating in community-supported agriculture or food co-ops is that they are delivered much more produce than they can eat in a timely fashion. One member of a local farm association admitted, "Ordinarily, I would never eat turnips. I managed to go 30 years without buying one. But now every winter I'm faced with a two-month supply, not to mention the kale, collards, and flat-leaf Italian parsley that sit in my refrigerator, slowly wilting, filling me with guilt every time I reach past them for the milk."[24]

Not only is this wasteful, it's not even all that healthy. The nutrient content of veggies that have been sitting around for a few days can deteriorate to the point that eating canned would be better—at least from a nutritional standpoint. That's why one group of scientists concluded "that exclusive recommendations of fresh produce ignore the nutrient benefits of canned and frozen products."[25]

Eating healthy means more than eating just a few kinds of vegetables. You won't do your body any favors with a broccoli-only diet. Nutritionists tell us that a key component of a healthy diet is the diversity of the foods we eat. For example, a recent paper in *The Journal of Nutrition* indicated, "The recommendation to eat diverse types of foodstuffs is an internationally accepted recommendation for a healthy diet."[26]

A person who restricts his diet to only those things grown locally is restricting diversity in his diet—especially in the winter. Wander around almost any supermarket in almost any town in America almost any time of year, and the diversity

and abundance of fruits and vegetables are absolutely astounding. Vidalia onions from Georgia, oranges from Florida,
California lettuce, sweet corn from Iowa, mangoes, bananas,
and jalapeños from south of the border—if you live in the
right location, you *might* have access to such a cornucopia a
few weeks or months out of the year, but Walmart offers it to
us every day. Fifty to a hundred years ago, the available transportation and storage technologies required people to eat a lot
more local food. Yet, despite weighing a bit less, people weren't
healthier then. One reason, among many, is that our great-
grandparents lacked the diversity of the diet we enjoy when we
eat food from places from beyond our backyard.

THE FIFTH TENET OF LOCAVORISM: LOCAL FOODS TASTE BETTER

At certain times of the year and with certain veggies, fresh
local food can taste better than nonfresh imported food. But
that's no argument for government promotion of local foods.
Dinner at a three-star Michelin-rated restaurant tastes better
than pizza from Domino's. Yet I don't see a campaign to limit
meals to only those cooked by chefs who can rate a star. Few
reasonable people would argue that governments should subsidize top chefs. Locavores must hold the rest of us in pretty
poor esteem if we can't be presumed to know what tastes best.

While I'll grant the locavores their premise at certain times
of the year, the same isn't true all year round. Maybe local
winter squash, broad beans, and beets do taste better than
the ones grown farther away. But in the winter, the real choice
isn't between local or imported winter squash. With abundant trade, we can choose between local winter squash and

imported peaches, oranges, and tomatoes. You'll have a hard time convincing me, and the vast majority of other Americans, that in February winter squash will taste better than peaches, oranges, tomatoes, and lettuce brought up from Florida or South America. No one comparing apples to oranges would win even a high school debate, and yet the locavores hope to convert us by hiding the fact that they're comparing winter squash to summer sweet corn.

Fresher often tastes better, but local food does not axiomatically mean fresh food. The fish in the city center of Paris is typically considered fresher than what can be found even a few miles in from the coast. (It goes to show the drawing power of a big consumer market.) Even in the landlocked Midwest, the best seafood restaurants fly in fresh catch. Want fresh peaches from Georgia or South Carolina? Thanks to the Internet and FedEx, you can have them sitting on your doorstep the next day.[27] That's as fresh as you'll find at any farmers' market.

THE SIXTH TENET OF LOCAVORISM:
LOCAL FOODS ARE FOR THE KIDS

Almost anything can be justified if it's for the kids. Alice Waters wants us to invest in local foods for schools because "[c]ertain things are just too important to compromise on. I just cannot compromise on what children are fed in school . . . I cannot compromise on the purity and nutrition and tastiness of the food. Children should have the very best of everything."[28] It is interesting that Waters says she can't compromise for *my* kids, because I have to do it every day.

The reality is that even Bill Gates has to compromise. Nobody's income is unlimited. Our wants will always outstrip our capacity to supply them. That's a fact of life—one that has escaped Waters, who apparently presumes schools have budgets more akin to those of the sheikhs of Saudi Arabia than administrators in the Bronx. Schools face real budget constraints that force real trade-offs. If schools are forced to spend their budgets on more local foods, that means less money spent on healthy foods. A school can try to maximize nutrient intake for a given budget or it can maximize the amount of local food. The outcomes are not the same. Personally, I'd rather schools tried to maximize what my kids learned.

Why should we promote policies that take fruits and veggies, which are often the farmers' most highly valued commodities, and give them to the people who value them least? Most kids don't want veggies—local or not. I'm all for local schools thinking about creative and cost-effective ways to improve the quality of schoolchildren's diets, but simply requiring schools to buy things kids won't eat isn't helpful. Even if we could increase local food consumption through subsidies, that doesn't make it an economically efficient thing to do. Giving out local food to seven-year-olds is about as smart as handing out T-bone steaks at a PETA convention.

When all is said and done, the locavores don't have a truly compelling answer for why we should all eat local. I once cowrote an article with the tongue-in-cheek subtitle "Why Pineapples Shouldn't Be Grown in North Dakota," and received a retort from a locavore saying they never claimed that pineapples should be grown there. "So, what do you claim?" I asked. The answer, in not so many words, was that people in North Dakota shouldn't *want* pineapples. There we have it.

Locavores don't care what we want. It's all about what *they* want us to eat. The food elite want us to become something different from what we are. Unless, ironically, a North Dakotan can export himself to where pineapples are local.

In the end, the locavores want to supplant our wants and desires with theirs. Art Salatin, a local-food advocate and farmer, says, "We have to battle the idea that you can have anything you want any time you want it."[29] This is an interesting position for someone in the business of trying to please the customer. At least Salatin's cause is a personal one that doesn't entail government regulation. That's not the same as Pollan, who says: "People will have to relearn what it means to eat according to the seasons."[30] Our modern economy allows us to eat tasty fruit from South America in the off-season, but if Pollan's crusade to have us eat more local food is to succeed, then our preferences must change. In fact, he argues, "[a]ll we need to do is empower individuals with the right philosophy and the right information to opt out [of the industrial food chain] en masse."[31] Apparently, according to Pollan, whatever philosophies are currently guiding food decisions are wrong.

Local-food advocates tend to be the exact same folk who promote cultural diversity and relish their adventures to Europe and Southeast Asia. Pollan himself traversed the United States to make the case that we should stay at home to eat. Locavores use populist language to extol the virtues of local producers, all the while smugly praising exotic foreign foods and spending their vacation money elsewhere. Despite the populist overtones, locavorism is but another form of elitism—even though locavores can't bring themselves to admit it. We have one group of people claiming to know better than the rest of us where we should get our food, all the while exempting themselves from the same requirements. The irony is lost on the locavores, who, in one breath, claim it is a myth that "eating local (organic) food is elitist," while, in the next, encourage fellow locavores to "continue enjoying your delicious Green Zebra and Brandywine tomatoes with a little bit of extra virgin olive oil, homegrown basil, and sea salt without the slightest bit of guilt."[32]

Most of us have more to worry about than the distance zucchinis travel. If your goal is to minimize nutrient loss, buy frozen veggies. If your goal is to feed kids cheaply and more nutritiously, buy canned. If you want to minimize your carbon footprint, ride your bike and stay off airplanes. If you want to help local farmers, give them a donation or fight to remove the trade barriers that keep them from selling to consumers in other countries. If you want to ingest fewer pesticides, pay more for the organic. Local has nothing to do with it.

The locavores have taken a movement designed to satisfy the whims of foodies with time and money on their hands and turned it into a national cause with the growing support of politicians. I happily buy from folks I know at the farmers' market when a ripe tomato is needed for fresh salsa, but I can't possibly imagine why everyone has to. As the award-winning author Charles Mann put it in a *New York Times* interview, "If your concern is to produce the maximum amount of food possible for the lowest cost, which is a serious concern around the world for people who aren't middle-class foodies like me, [local food] seems like a crazy luxury. It doesn't make sense for my aesthetic preference to be elevated to a moral imperative."[33]

10

THE FUTURE OF FOOD

A couple years ago, I had the pleasure of giving a speech about the economics of the U.S. beef industry to a group of visiting Vietnamese businessmen and government officials. In the course of my talk, I showed two maps of the United States, to illustrate where cattle are raised and subsequently processed. At the conclusion, I received several questions from members of the audience wanting to know *why* cattle were raised all over the country and then shipped to feedlots and packing plants primarily located in a relatively small geographical area spanning the panhandles of Texas and Oklahoma and extending up into Kansas and Nebraska. I answered with all the economic rationalizations I could muster, citing weather, corn prices, transportation costs, and economies of scale.

My answers were apparently unsatisfactory, because the queries persisted. Finally, the interpreter—a Vietnamese native who had spent several years living in the States—stopped me, switched to English, and said, "You have to remember they live in a Communist country. The government decides where firms locate." Suddenly I realized that my listeners had no conception of how this sort of efficient outcome could have resulted from the free interaction of ranchers, feedlot owners, and agribusinesses all trying to make a living satisfying consumers' preferences for beef. These Vietnamese businessmen wanted to know *who* decided. I had to reveal that we have no cow czar, but I could tell they were still a bit bewildered. It was beyond their worldview to consider the possibility of a prosperous and successful agricultural economy running on the backs of unplanned markets coordinating people's free choices.

This experience opened my eyes to how strongly our beliefs and perceptions are influenced by the culture in which we live, and how, over the course of several decades, entire worldviews can change. The free-market apologist and queen of individualism, Ayn Rand, immigrated to the United States in the 1920s as the Communists began overtaking Russia. She left behind a sister for whom she had fond memories. In 1973, after years of believing her sister had died, Rand learned that she was in fact alive and well, and arranged for her to visit New York. Rand, who fondly remembered her older sister as spirited and enamored of Western fashion, was dismayed at the changed woman who arrived. Rand's sister, who by now had lived a near-lifetime under Soviet rule, "disliked American conveniences, which left her nothing to do all day; she preferred her old routine of waiting in the food lines and gossiping with her friends." She was also overwhelmed at the variety offered by American grocery stores. Like the Vietnamese I addressed,

Rand's sister was perplexed by choice. She "went into a store to buy a tube of toothpaste and found herself overwhelmed by the number of brands and sizes and became angry when a clerk wasn't willing to help her choose."[1] Two sisters who had once rejoiced in their similarities found they had little in common, in no small part because such divergent cultures and ideas had permeated their thinking.

One of the ways we make it through life with even the tiniest sliver of sanity is by adopting ways of thinking to give order to the capriciousness that surrounds us. Some folks spend years trying to develop a view on life, a worldview, to make sense of their happenings. But no matter how reflective or thoughtful, we are all unconsciously influenced by the culture in which we live. Yet ideologies and worldviews have real implications for how we interact in our world and the outcomes we can expect to enjoy.

I write about the food police because they have been spectacularly effective in influencing how our culture thinks about food. A generation of Americans far removed from the farm is one that lacks the tangible hands-on experience to judge for themselves the veracity of claims about modern food. Missing is the connection required to know whether a worldview should be kept or checked at the door.

It's hard to know what to believe about food, and sometimes it's easier to take our cues from those around us than to think deeply about controversial issues. I write to those who, perhaps unconsciously, have absorbed the messages of celebrity chefs, urban journalists, and daytime talk show hosts and who have adopted certain attitudes toward food and agriculture without previously having stopped to think. Charles

Mackay, writing about the *Madness of Crowds,* argued that "[m]en, it has been well said, think in herds; it will be seen that they go mad in herds, while they only recover their senses slowly, one by one."[2] Sometimes it is useful to stop, look up, see where the crowd is heading, and ask if you still want to be on board.

I grant that among the folks I've called the food police there are many intelligent, well-meaning people who truly care about our health and the environment. Here's the thing. I care about these, too. It isn't a question of who is more compassionate; it is a question of *how* we can best achieve the things we all want. It is a question of costs. It is a question of who is best suited to make decisions about the tough trade-offs. It is a question of the kind of future we hope to create. It is a question of worldview.

A danger embedded in the worldview of the progressive food elite is romanticism with nature and the past. The food progressives reject modern, technologically advanced food for older, slower, more "natural" food. The answer for the future of food, in their assessment, lies in the past. This mind-set is destructive because how we view our future affects not only who we are but what we eventually will be. We travel to Athens and Rome to admire who they *were* and what they *did,* but seldom do we think they represent what we should become. According to the economist Deirdre McCloskey, the modern Western world is incredibly wealthy today not so much because of the industrial revolution and capitalism per se but because of changes in how we came to *think* about commercialism, private property, innovation, and trade.[3] Aspirations are more than wishes—they affect who we become.

A worldview that celebrates naturalism in food as a core tenet is one inherently hostile to innovation, growth, and progress, even if those things reduce poverty and bring about improved food safety, quality, health, and environmental outcomes. The food police type away on their iPads and Tweet to followers. Yet, ironically, they have rejected biotechnology, irradiation, cloning, synthetic fertilizers, and all kinds of technology, mechanization, and innovative processing in agriculture. Each of these developments admittedly has risks and downsides, but all can make food more abundant, more convenient, more consistent, sometimes less reliant on input use (especially labor), and sometimes more environmentally friendly.

The policies of the food police will usher in a more stagnant, less dynamic world, and will breed a generation of children unwilling or unable to imagine how to improve their diets through mathematics, chemistry, biology, and engineering. I don't want a future where my kids brood over the oldest heirloom tomatoes but one where they dream about how to get corn to produce its own fertilizer, feeding the world's hungry with higher-yielding, more nutritious crops, and developing space-age technologies that make tasty food at the push of a button. Those dreams may never materialize, but they are destined to fail if we never try.

It is almost unfathomable to imagine how our current world would look had previous generations adopted the thoughts about food that now pervade our culture. Over the last century there has been a threefold increase in the world's population. The spike occurred in large part because innovators figured out ways to get more food from the same planet we've always occupied. According to Vaclav Smil, author of *Enriching the Earth*, the key catalyst was the discovery of synthetic fertilizers. That development, Smil argues, has been "of greater fundamental importance to the modern world than the invention of the airplane, nuclear energy, space flight, or television."[4] However much the food police love organic, chances are that without "unnatural," synthetic fertilizers, they wouldn't be here today—nor in all likelihood would you or I. So, when we promote or even subconsciously adopt a worldview hostile to innovation in food, we are closing off a bright and optimistic future in favor of remnants of the past.

The food police not only hail naturalism but promote utopianism—denying the reality of the tough trade-offs that face us in life. They have taken boutique food made for the rich and fancifully imagined a world where the poor can afford it, too. It is a beautiful vision, but one at odds with reality. However much they wish it were different, organic, local, slow crops are typically lower-yielding crops. Even if we didn't have the distorting effects of farm policies and even if all the negative and positive externalities were fully internalized (in both organic and nonorganic production systems), it would probably remain the case that the fashionable food would cost more. What I'm at least willing to admit, and what the food police seem to deny, is that there is a trade-off between an abundant food supply that can feed a growing population and a higher-quality, boutique food supply that can suit the fancy of foodies. It doesn't in the least bother me that there are niche markets catering to the whims of a few, but when the food police act as if national policies promoting fancy food are a net plus, it is a triumph of ideology over reason.

The French have a saying describing the *gauche caviar*—the political leftist who says he wants to help the poor, just not live next to them: *"Il a des idées à gauche, mais le portefeuille à droite"*—which roughly means the person has his ideas in his left pocket but keeps his wallet in his right. The food elite are brimming with ideas on how to make over the food system, imagining that their plans will produce flowering fields, wealthy communities, and thin waists. But rarely do they take out their wallets to cover the costs. And even when they pay the higher costs of local and organic for themselves, seldom do they consider the effects of their policy ideas on others who have different wants, needs, and budget constraints.

The food elite's worldview causes them to see large farms, agribusinesses, and restaurants as evil conspirators who will stop at nothing to earn an extra buck. Never mind that none of these folks can earn a living unless we are willing to buy what they offer. Big Food, in the mind of the food police, is an omnipotent opinion maker with powers akin to those of Darth Vader. Meanwhile, I see meatpackers and fast-food restaurants that are one food safety scare away from bankruptcy; agribusinesses constantly trying to innovate to make food tastier, less expensive, and more nutritious than that of their competitor. Because, despite all their advertising, you and I are the ones with the wallets, and thus we control what Big Food can actually sell.

Where the food police see a benevolent government with the ability to rein in the power of Big Food, I see the development of arbitrary rules breeding the worst kind of crony capitalism. Ironically, the food police can see this, too, in our current food system. They blame farm policy, fertilizer subsidies, and agribusiness lobbyists for everything from obesity to topsoil degradation. Rather than accept the obvious reality that a government willing to dole out favors and subsidies is one that invites corruption and, through its actions, creates externalities and unintended consequences, the food police see a savior for our food ills. There is no perfect government except in textbooks.

That government is imperfect does not imply that markets are free from error. Yet the ideology of the food police is hostile to the good that even imperfect markets provide. Markets are the enemy of the food police because they undermine the food police's power to enact their preferred plans for us. The truth is that we all have different preferences and unique knowledge that cannot be rationally subsumed in a grand food plan. Markets decentralize power and let us each satisfy our desires in unique ways. McDonald's, ADM, and Monsanto seem big and all-powerful—until one realizes that they must compete with Wendy's, Cargill, and DuPont for the dollars we spend. These corporations may not provide a lot of organic veggies now, but that doesn't mean they wouldn't if folks were willing to pay for them. That many of us are unable or unwilling to pay the higher price is anathema to the food elite, who seem to think that we all should eat as they do. The food police see themselves as fighting on our behalf to stave off the power of Big Food, but it is raw condescension to provide help when it hasn't been asked for.

Allowing us to buy and sell freely and as we wish denies the food elite their role as holders of special knowledge. My call for calm in the rising seas of food regulation is a plea for humility from those who think they know how to guide our nation's food choices. I honestly don't know whether, all things considered, organic is better for the environment. What I do know is that most people aren't willing to pay the price it takes to make food that way. I don't know with absolute certainty whether biotechnology has made farmers more profitable or helped the environment. What I do know is that farmers all over America, of their own volition, choose to pay a little more to plant it in the fields next to their homes on their family-owned farms. I don't know precisely why Americans—and people all over the world, for that matter—are fatter than they once were. What I do know is that most of us eat better and live longer lives than our ancestors ever dreamed possible. The food police, confident in their own academic insights and neat explanations, boldly tell us all how to eat, blind to their own ignorance of the complexity of the world that surrounds them.

The prevailing wisdom about food that permeates popular culture has been formed by muckraking journalists, cookbook authors, and celebrity chefs. These folks know a lot about how to write compelling exposés and make tasty food using the finest ingredients, but are they really qualified to step into the public arena with pronouncements about how we all should eat? When they claim to know more about how to run a farm than the flesh-and-blood people who work it, surely something is amiss. Deciding what to eat can be difficult enough in its own right without the morally laden subtexts that have been thrown into the matter by the food elite.

Make no mistake about it, you are in the middle of a food fight. The problem is that many of us have joined sides without even realizing a battle is at hand. As the economist Walter

Williams put it, "The anti-obesity movement is simply another step down the road to serfdom and, what's worse, Americans are voluntarily assisting the nation's tyrants."[5] There will no doubt be detractors who deny the existence or seriousness of the food fight. After all, where are the wounded with battle scars? Fair enough. Americans do not currently face an existential threat to their food supply (that is, unless you believe some of the food progressives' apocalyptic prophecies). But we should at least be aware that there is an ideological movement afoot that has plans for farm and fork.

Many have bought hook, line, and sinker into the local, organic, and natural food movements without carefully considering where they are being led—or in many cases pushed. It has been accepted as gospel in many quarters that genetically modified foods are one of the greatest evils to enter the food supply. We are told that local-food systems will save the economy, the environment, and our health. But sometimes our sacred cows must be slaughtered to get closer to the truth.

No doubt my perspectives will earn me the label of shill for corporate farms and agribusinesses, so it is important that I not be misinterpreted. On Saturday mornings, I like nothing more than to drive my children down to the local farmers' market in my hybrid SUV. (Yes, I drive a hybrid.) I love to shop at Whole Foods and Dean & Deluca. Walk in our house any weekday evening and you're likely to find us watching the Food Network. My family spent most of last year in Paris, in part because we are fascinated with how the French relate to food.

I do these things not because I am convinced of a moral imperative to eat or act a certain way with regard to food, but because I want to and because I can afford to. But it isn't always this way. When I don't find what I want at the farmers' market or when the budget gets tight at the end of the month, I have no qualms about stopping by Walmart on the way home (and truth be told, most of our family's food is bought there).

Knowledge about the reality of agriculture coupled with some rational thinking about food empowers one to eat with freedom and without guilt. Reason should guide food choice, and responsible choices will not necessarily be found in the dictates of the food elite. Responsibility lies not just in thinking about the environment, freshness, or animal welfare, but also in spending our money and resources wisely.

Many of the world's most pressing challenges—from feeding the hungry to protecting the environment—require a look forward rather than backward. We should be cautiously optimistic about the role of technological development in food and be leery of the ideologically motivated scare tactics used to denigrate food technologies. If we are concerned about food prices and feeding the world's hungry, we must be willing to rationally weigh the risks of, and perhaps even promote, technological change in food and agriculture. There are many exciting—and scary—advances on the horizon that will require serious thought and discussion. One benefit of ideology is that it helps us form quick judgments in a complex and rapidly changing world. But an ideology that blanket-rejects any "unnatural" food is one that will ultimately doom us to poverty.

Using fewer pesticides, eating more veggies, or supporting a local farmer can all be good things in their own right. But these choices involve tough trade-offs. The food elite throw caution to the wind and appear willing to make the choices for us. I'm willing to make a tough trade-off. I'd rather live in a world where I'm tempted by fast-food restaurants and weigh a few more pounds than one where I have no choice at all.

It is time to dispense with the food guilt. We seem to have forgotten the achievements of our efficient and safe food system. It is not a perfect system but it is one that has led to the greatest prosperity ever witnessed in human history and has allowed us to feed the world. I am rationally optimistic about the future of food. My optimism lies not in the ability of the food elite to usher in an optimal system of taxes, subsidies, and mandates. My hope is in the dynamism and innovativeness of farmers and grocers trying to make a buck by selling us things we don't yet know we even want.

Let's get out of their way.

ACKNOWLEDGMENTS

This book was written while I was on sabbatical in Paris. My wife, Christy, moved halfway across the world and temporarily homeschooled our children to make the sabbatical, and this book, possible. She read countless drafts of this book, and her suggestions made it immeasurably better. When I fretted about the potential negative side effects of writing this sort of book on my academic career, she helped remind me of the importance of what I was doing. She and our children, Jackson and Harrison, are important reminders that there is more to life than the social science citation index. Nothing I can say can adequately express my gratitude to her. Her dear friends Amy, Jessica, and Shayna, and her mother, Marcia, provided valuable insight into what regular mothers, who don't continually brood over food policy, think about the issues raised in this book.

My father, Raymond Lusk, helped me better understand school-lunch policies, and he is most specifically to blame for my firsthand knowledge of production agriculture. For ten years of my youth, he bought my siblings, Keri, Kay Lynn, Thad, and me a slew of farm animals to watch after for FFA and 4-H projects, and he made sure the critters didn't starve when we were too lazy to roll out of bed in the mornings. Somehow one of his farming buddies always seemed to have a job waiting for us when school was out of session. My mother taught me that the kitchen wasn't a woman's domain, and instilled a lifelong love of cooking that deeply affected how I think about food. Whatever advantages I enjoyed in life, I owe to Mom and Dad's self-control and willingness to sacrifice their present for my future.

Even though my family doesn't farm or process food for a living, my perspectives on food and agriculture have been formed by a lifelong connection with many wonderful families who do. I am grateful to the Baldwin, Burris, and Brackeen families, who, early in life, gave me the jobs that first taught me about food and agriculture. Many kind county extension agents and high-school ag teachers also influenced my thinking on production agriculture. Max Miller and Leslie Thompson were exceptional mentors during my undergraduate years, and they taught me a great deal about food science and technology.

I have the privilege of working in an outstanding academic community of agricultural economists, and I am thankful to Jon Biermacher, Keith Coble, Eric DeVuyst, Andreas Drichoutis, Fabrice Etilé, Darren Hudson, John Lee, Stéphan Marette, Lanier Nalley, Rudy Nayga, and Steve Sexton, among others who I have probably forgotten to mention, who read various chapters and provided valuable feedback and criticism. I am also grateful for the feedback I received from people who attended the seminars I presented on behavioral economics at the University of Wisconsin; Texas A&M University; a preconference workshop on behavioral and experimental economics, food, and health at the Agricultural and Applied Economics Association; and at the French National Institute for Agricultural Research. There has probably been no bigger influence on my thinking about food, politics, and economics than Bailey Norwood. For all his offbeat questions, philosophizing, and good-natured arguments, I will be forever grateful. Without his constant encouragement, I surely would have given up on this project.

Thanks to Stéphan Marette and the French National Institute for Agricultural Research (INRA) for being gracious sabbatical hosts and to the Organisation for Economic Cooperation and Development (OECD) for additional financial support. Oklahoma State University (OSU) kindly gave me the time away, and for that, I am grateful. The views expressed in this book are entirely mine and are not those of INRA, OECD, or OSU.

I will probably never know what possessed my agent, Mel Berger, to take on the project of someone virtually unknown outside the circle of a couple thousand academic economists. But I am grateful to him for taking the risk and providing feedback on early drafts of this work. Finally, thanks to Sean Desmond at Crown Forum for his excellent editorial advice and to Paul Lamb and the marketing team for their support.

NOTES

Author's note: All URLs provided herein were last accessed on April 5, 2012.

CHAPTER 1: A SKEPTICAL FOODIE

1. The FDA required mandatory labeling of trans fats in 2003, and cities such as New York and Philadelphia subsequently banned the use of trans fats in restaurants. Earlier, in 2002, Kraft reformulated Oreos in response to a lawsuit filed by a San Francisco lawyer.

2. In a testimony to the unintended consequences, it appears that McDonald's found a way around the ban by selling the toys rather than giving them away. Now, in San Francisco, the only way to get a Happy Meal is to buy a toy. The quote is from S. Bernstein, "San Francisco Bans Happy Meals," *Los Angeles Times,* November 2, 2010.

3. michellemalkin.com/2010/11/03/san-francisco-bans-happy-meals/.

4. Barry Popkin, *The World Is Fat* (New York: Avery/Penguin USA, 2008), pp. 162–63. The 100 percent figure comes from Popkin's statement, "If a one-dollar can of Pepsi or Coke costs two dollars instead, we would find—as we have for cigarettes—a significant reduction in soft drink consumption."

5. From Bill Moyers's TV show on November 28, 2008, www.pbs.org/moyers/journal/11282008/transcript1.html.

6. www.agweb.com/article/house_approves_legislation_that_puts_gipsa_on_hold/.

7. Mark Bittman, "Bad Food? Tax It, and Subsidize Vegetables," *New York Times,* July 23, 2011. A printed version appeared on July 24, 2011, on page SR1 of the New York edition with the headline "Bad Food? Tax It."

8. For a fuller discussion on this issue, see my coauthored book

Compassion by the Pound: The Economics of Farm Animal Welfare (New York: Oxford University Press, 2011).

9. Kim Severson, *Spoon Fed* (New York: Riverhead Books/Penguin USA, 2010), pp. 67–68.

10. M. Muskal, "Obama Signs Child Nutrition Bill, Championed by the First Lady," *Los Angeles Times,* December 13, 2010.

11. www.cnsnews.com/news/article/nyc-mayor-bloomberg -government-s-highest-duty-push-healthy-foods.

12. *Health Affairs* 29, no. 3 (2010): 342.

13. Powell et al., *Journal of the American Dietetic Association* 110 (2010): 847.

14. K. Dun Gifford quoted in Craig Lambert, "The Way We Eat Now," *Harvard Magazine,* May/June 2004, p. 99.

15. For example, see Michael Pollan, *The Omnivore's Dilemma: A Natural History of Four Meals* (New York: Penguin, 2006).

16. Jonah Goldberg, *Liberal Fascism* (New York: Doubleday, 2007).

17. Ludwig von Mises, *Human Action* (Auburn, AL: Ludwig von Mises Institute, 2008; originally published in 1954), p. 10.

18. D. Cutler, E. Glaeser, and J. Shapiro, "Why Have Americans Become More Obese," *Journal of Economic Perspectives* 17 (2003): 93–118.

19. www.ers.usda.gov/Briefing/CPIFoodAndExpenditures/Data/Table _97/2008table97.htm.

20. www.ers.usda.gov/Briefing/CPIFoodAndExpenditures/Data/ Expenditures_tables/table8.htm.

21. Ibid.

22. Statistics on number of restaurants and female labor force participation are in: I. Rashad, M. Grossman, and S. Chou, "An Economic Estimation of Body Mass Index and Obesity in Adults," *Eastern Economic Journal* 32 (2006): 133–48.

23. The statistic is the percentage change in the average per capita consumption over the years 1970 to 1979 to the average per capita consumption over the years 2000 to 2009. See www.ers .usda.gov/data/foodconsumption/FoodAvailspreadsheets .htm#vegtot.

24. The statistic is the percentage change in the average per capita

consumption over the years 1970 to 1979 to the average per capita consumption over the years 2000–2009. http://www.ers .usda.gov/data/foodconsumption/FoodAvailspreadsheets .htm#frtot.

25. Centers for Disease Control and Prevention, "Preliminary FoodNet Data on the Incidence of Infection with Pathogens Transmitted Commonly Through Food—10 states, 2009," *Morbidity and Mortality Weekly Report* 59 (2010): 418–22. The 25 percent figure comes from calculating the total incidences for all bacteria reported for 2009 in Table 1 and using the data reported below Figure 1 to calculate total 1996–98 incidence rates, and then computing the percentage change in total incidence from 1996–98 to 2009.

26. www.cdc.gov/nchs/data/hus/hus10.pdf#022; www.cdc.gov/nchs/ data/nvsr/nvsr59/nvsr59_04.pdf; www.cdc.gov/nchs/data/nvsr/ nvsr60/nvsr60_04.pdf.

27. The last twenty years refers to the period 1990–2010, usda. mannlib.cornell.edu/usda/current/htrcp/htrcp-04-26-2011.pdf; more evidence of growth in agricultural productivity can be found at www.ers.usda.gov/Data/AgProductivity/.

28. Half as much active ingredient per acre planted; at least up to 1995, which was the last year given in the report, www.ers.usda .gov/publications/arei/ah712/AH7123-2.pdf.

29. The corn statistic is derived from figure 8 and the quote is from page 1 in J. Fernandez-Cornejo et al., "Assessing Recent Trends in Pesticide Use in U.S. Agriculture," selected paper presented at the 2009 Meetings of the AAEA, Milwaukee, Wisconsin, July 2009.

30. Statistics are computed from USDA data available at www.ers .usda.gov/Briefing/FarmIncome/Data/Constant-dollar -table.xls.

31. www.ers.usda.gov/briefing/wellbeing/farmhouseincome.htm; www.ers.usda.gov/Briefing/WellBeing/Gallery/ PerCapitaDisposablePersonalIncome3483.htm.

32. "Past two decades" refers to the period 1992–2009, www.ers.usda .gov/amberwaves/june10/Indicators/InTheLongRun.htm.

33. Chapter 2, "Profiling Food Consumption in America," in

Agricultural Fact Book, 2001–2002, U.S. Department of
Agriculture, Office of Communications, March 2003.

CHAPTER 2: THE PRICE OF PIETY

1. CSPI's Nutrition Action Healthletter, July/August 1998, U.S.
   edition, www.cspinet.org/nah/7_98eat.htm.
2. Michael Pollan, *The Omnivore's Dilemma: A National History of
   Four Meals* (New York: Penguin, 2006), p. 2.
3. Stated in a TED talk entitled "Mark Bittman on What's Wrong
   with What We Eat," www.ted.com/talks/mark_bittman_on_what
   _s_wrong_with_what_we_eat.html.
4. An estimated twenty excess cancer deaths are caused each year
   in the United States from the use of agricultural pesticides;
   three hundred deaths are from drowning in the bathtub. See
   Bjørn Lomborg, *The Skeptical Environmentalist* (New York:
   Cambridge University Press, 2001), p. 245. There are more than
   thirty thousand deaths from automobile accidents each year. See
   www.cdc.gov/nchs/data/nvsr/nvsr59/nvsr59_04.pdf.
5. The statements in this paragraph are based on the findings from
   the following studies: J. F. M. Swinnen, P. Squicciarini, and
   T. Vandemoortele, "The Food Crisis, Mass Media, and the
   Political Economy of Policy Analysis and Communication,"
   European Review of Agricultural Economics 38 (2001): 409–26;
   S. Mullainathan and A. Sheifer, "The Market for News," *American
   Economic Review* 95 (2005): 1031–53; J. J. McCluskey and J. F. M.
   Swinnen, "Political Economy of the Media and Consumer
   Perceptions of Biotechnology," *American Journal of Agricultural
   Economics* 86 (2004): 1230–37.
6. See R. S. Taylor et al., "Dietary Salt for the Prevention of Cardio-
   vascular Disease: A Meta-Analysis of Randomized Controlled
   Trials," *American Journal of Hypertension* 24 (2011): 843–53;
   K. Stolarz-Skrzypek et al., "Fatal and Nonfatal Outcomes,
   Incidence of Hypertension, and Blood Pressure Changes in
   Relation to Urinary Sodium Excretion," *Journal of the American
   Medical Association* 305 (2011): 1777–85; www.scientificamerican
   .com/article.cfm?id=its-time-to-end-the-war-on-salt.

7. intelligencesquaredus.org/wp-content/uploads/Organic-041310 .pdf.
8. Pollan, *The Omnivore's Dilemma*, p. 318.
9. The quote is from George Naylor, in ibid., p. 55.
10. Stan Cox, *Sick Planet: Corporate Food and Medicine* (London: Pluto, 2008).
11. M. Nestle, *Food Politics* (Berkeley: University of California Press, 2007), p. 4.
12. www.thenation.com/article/163403/food-movement-its-power -and-possibilities.
13. A. Lotti, *Agricultural and Human Values* 27 (2010): 71.
14. press-pubs.uchicago.edu/founders/documents/amendI_speechs2 .html.
15. F. A. Hayek, *The Road to Serfdom* (Chicago: University of Chicago Press, 2007), pp. 130–31.
16. See A. Bourdain, *Medium Raw* (New York: HarperCollins, 2010), pp. 62–63.
17. The quote is by Rahm Emanuel, made in November 2008.
18. A. Coleman-Jensen et al., "Household Food Security in the United States in 2010," United States Department of Agriculture, Economic Research Service, Economic Research Report Number 125, September 2011.
19. John Steinbeck, *Travels with Charley: In Search of America* (New York: Penguin Books, 1980), p. 83.
20. Pollan, *The Omnivore's Dilemma*, pp. 63 and 64.
21. The statement was made at a November 13, 2002, meeting of the American Public Health Association.
22. "Food Crisis," *New York Times*, February 24, 2011. The article appeared in print on February 25, 2011, on page A26 of the New York edition.
23. Mark Bittman, "Don't End Agricultural Subsidies. Fix Them," *New York Times*, March 2, 2011. To be clear, these authors argue that food is too cheap because of negative externalities associated with modern food production. I address this issue in chapter 7.
24. V. Shiva, "Resisting the Corporate Theft of Seeds," *The Nation*, October 3, 2011.

25. J. M. Beddow, P. G. Pardey, and J. M. Alston, "The Shifting Global Patterns of Agricultural Productivity," *Choices* (2009, 4th quarter).

26. L. Aron, "Everything You Think You Know About the Collapse of the Soviet Union Is Wrong," *Foreign Policy,* July/August, 2011.

27. David Horowitz, *The Politics of Bad Faith* (New York: Touchstone Books, 2000).

28. George W. Breslauer, *Gorbachev and Yeltsin as Leaders* (New York: Cambridge University Press, 2002), p. 249.

29. Thomas Sowell, *Intellectuals and Society* (New York: Basic Books, 2009), p. 118.

30. www.nytimes.com/2011/11/11/opinion/the-inequality-map.html?_r=1.

31. www.economist.com/node/8450035.

CHAPTER 3: FROM COPS TO ROBBERS

1. F. A. Hayek, *The Road to Serfdom* (Chicago: University of Chicago Press, 2007), p. 57.

2. J. Thirsk, *Alternative Agriculture: A History from the Black Death to the Present Day* (Oxford: Oxford University Press, 1997).

3. Quotes are from "William Bradford, of Plymouth Plantation, 1620–1621," The Founders' Constitution, press-pubs.uchicago.edu/founders/documents/v1ch16s1.html.

4. Adam Smith, *An Inquiry into the Nature and Causes of the Wealth of Nations,* 1776, available online at www.econlib.org/library/Smith/smWN1.html.

5. J. E. McWilliams, *A Revolution in Eating* (New York: Columbia University Press, 2005), p. 320.

6. These data are from the U.S. Department of Agriculture. See also inventors.about.com/library/inventors/blfarm4.htm.

7. A fascinating account of this history is in D. Okrent, *Last Call: The Rise and Fall of Prohibition* (New York: Scribner, 2010).

8. www.econtalk.org/archives/2010/06/okrent_on_prohi.html.

9. See chap. 31 in Upton Sinclair's *The Jungle.*

10. This quote can be found in the preface of the edition of *The Jungle* published by Simon & Schuster in 2004.

11. C. E. Ward, "Assessing Competition in the U.S. Beef Packing Industry," *Choices* 25, no. 2 (2nd quarter): 2010.

12. A. M. Azzam and J. R. Schroeter, "The Tradeoff Between Oligopsony Power and Cost Efficiency in Horizontal Consolidation: An Example from Beef Packing," *American Journal of Agricultural Economics* 77 (1995): 825–36.

13. According to USDA data compiled by the Livestock Marketing Information Center: /www.lmic.info/. The statistics refer to the change in average monthly real prices in 1970 to the average monthly real prices in 2011.

14. J. M. Alston, M. A. Andersen, J. S. James, and P. G. Pardey, "The Economic Returns to U.S. Public Agricultural Research," *American Journal of Agricultural Economics* 93 (2011): 1257–77.

15. Ibid. W. E. Huffman and R. E. Evenson, "Do Formula or Competitive Grant Funds Have Greater Impacts on State Agricultural Productivity?" *American Journal of Agricultural Economics* 88 (2006): 783–98.

16. See, for example, C. B. Moss, "Valuing State-Level Funding for Research: Results for Florida," *Journal of Agricultural and Applied Economics* 38 (2006): 169–83. More generally, see J. M. Alston et al., *Persistence Pays: U.S. Agricultural Productivity Growth and the Benefits from Public R&D Spending* (New York: Springer, 2009), or J. M. Alston, G. W. Norton, and P. G. Pardey, *Science Under Scarcity* (Ithaca, NY: Cornell University Press, 1995; CAB International, 1998).

17. This calculation is in R. Roberts, *The Price of Everything* (Princeton, NJ: Princeton University Press, 2008).

18. Based on data reported at: www.ers.usda.gov/Data/Ag Productivity/table01.xls. The data ended in 2009, which is the endpoint I use to compare against 1950. The land use figure reported in the following sentence is also derived from the same data set.

19. www.law.cornell.edu/supct/html/historics/USSC_CR_0317_0111 _ZO.html.

20. See R. L. Paarlberg, *Food Politics: What Everyone Needs to Know* (New York: Oxford University Press, 2010), or J. Goldberg, *Liberal Fascism* (New York: Broadway Books, 2007).

21. D. Hannan, *The New Road to Serfdom: A Letter of Warning to America* (New York: HarperCollins, 2010), p. xvi.

22. S. Nasar, *Grand Pursuit: The Story of Economic Genius* (New York: Simon & Schuster, 2011), p. 24.

CHAPTER 4: ARE YOU SMART ENOUGH TO KNOW WHAT TO EAT?

1. B. Wansink and J. Sobal, "Mindless Eating: The 200 Daily Food Decisions We Overlook," *Environment and Behavior* 39 (2007): 106–23.

2. Statement made in a talk to the National Restaurant Association in Washington, D.C., on September 19, 2011, cnsnews.com/news/article/usda-secretary-we-must-create-appropriate-transition-what-americans-eat.

3. whitehouse.blogs.cnn.com/2011/09/15/the-never-ending-pasta-bowl-goes-healthy-with-the-first-lady/?iref=allsearch.

4. It is worth noting that this is a literature to which I, too, have contributed; e.g., see A. Arunachalam, S. R. Henneberry, J. L. Lusk, and F. B. Norwood, "An Empirical Investigation into the Excessive-Choice Effect," *American Journal of Agricultural Economics* 91 (2009): 810–25; D. Roberts, T. Boyer, and J. L. Lusk, "Preferences for Environmental Quality Under Uncertainty," *Ecological Economics* 66 (2008): 584–93; J. L. Lusk and K. O. Coble, "Risk Aversion in the Presence of Background Risk: Evidence from the Lab," in *Risk Aversion in Experiments*, J. C. Cox and G. W. Harrison, eds., *Research in Experimental Economics*, vol. 12 (Bingley, UK: Emerald, 2008). However, finding that people sometimes make mistakes does not axiomatically imply the need for paternalism. Stated differently, there are some behavioral economists who do not advocate paternalism.

5. D. R. Just, L. Mancino, and B. Wansink, "Could Behavioral Economics Help Improve Diet Quality for Nutrition Assistance Program Participants?" U.S. Department of Agriculture, Economic Research Service, Economic Research Report No. (ERR-43), June 2007, www.ers.usda.gov/publications/err43/.

6. C. Camerer, S. Issacharoff, G. Loewenstein, T. O'Donoghue, and

M. Rabin, "Asymmetric Paternalism," *University of Pennsylvania Law Review* 151 (2003): 1211.

7. G. A. Akerlof, "Procrastination and Obedience," *American Economic Review* 81 (1991): 2.

8. Richard H. Thaler and Cass R. Sunstein, *Nudge* (New Haven, CT: Yale University Press, 2008), p. 7.

9. Ibid., p. 19.

10. D. Ariely, "Dan Ariely Picks the Seven Most Powerful New Economists," Forbes.com, November 4, 2010, www.forbes.com/ 2010/11/01/dan-ariely-world-bank-opinions-powerful-people-10 -new-economists_slide.html.

11. Kahneman and Thaler concluded, "People do not always know what they will like; they often make systematic errors . . . and, as a result, fail to maximize their experienced utility," whereas Kagel et al. concluded, "Animals have behavioral repertoires that can be characterized as solutions to a constrained maximizing model." D. Kahneman and R. H. Thaler, "Anomalies: Utility Maximization and Experienced Utility," *Journal of Economic Perspectives* 20 (2006): 221–34; J. H. Kagel et al., "Demand Curves for Animal Consumers," *Quarterly Journal of Economics* 96 (1981): 1–16.

12. The quote is from www.fullerthaler.com/Default.aspx.

13. The quote was in reference to J. P. Morgan's "Undiscovered Managers Behavioral Growth Fund" (ticker symbol: UBRLX), for which Fuller and Thaler serve as sole advisers. The statement was made by Christopher Davis on June 1, 2010.

14. A. Santoni and A. R. Kelshiker, "Behavioral Finance: An Analysis of the Performance of Behavioral Finance Funds," *Journal of Index Investing* 1 (2010): 56.

15. C. Wright, V. Boney, and P. Banerjee, "Behavioral Finance: Are the Disciples Profiting from the Doctrine?" (August 22, 2006), p. 33, available at SSRN ssrn.com/abstract=930400.

16. D. Ariely, *Predictably Irrational: The Hidden Forces That Shape Our Decisions* (New York: HarperCollins, 2008), p. 62.

17. Ibid., p. 118. He also offers a compromise "if mandatory health checkups won't be accepted by the public," by suggesting the use of self-imposed deadlines and penalties for procrastinators.

18. G. Loewenstein and P. Ubel, "Economics Behaving Badly," *New York Times*, July 15, 2010, p. A31.

19. The culprit for the decreasing puberty age is likely related to the fact that adolescents are now better fed and weigh more than they once did. For example, see the *New York Times*, www.nytimes.com/2005/03/08/health/08real.html.

20. Data on prices of organic versus conventional milk prices can be found in K. Brooks and J. L. Lusk, "Stated and Revealed Preferences for Organic and Cloned Milk: Combining Choice Experiment and Scanner Data," *American Journal of Agricultural Economics* 92 (2010): 1229–41.

21. L. Stewart, *Implanting Beef Cattle*, University of Georgia Cooperative Extension Service, Bulletin 1302, February 2010, www.caes.uga.edu/publications/pubDetail.cfm?pk_id=7430.

22. The propensity to overweight low-probability events was one of the cornerstones of prospect theory, which was introduced by the psychologists Daniel Kahneman and Amos Tversky in 1979. This theory is sometimes credited with the birth of modern-day behavioral economics. See "Prospect Theory: An Analysis of Decision under Risk," *Econometrica* 47 (1979): 263–92.

23. V. K. Smith and E. M. Moore, "Behavioral Economics and Benefit Cost Analysis," *Environmental and Resource Economics* 46 (2010): 232.

24. L. Mises, *Human Action* (Auburn, AL: Ludwig von Mises Institute, 2008; originally published in 1954), p. 19.

25. Thomas Sowell, *Intellectuals and Society* (New York: Basic Books, 2009), pp. 13–14.

26. S. D. Levitt, J. A. List, and D. Reiley, "What Happens in the Field Stays in the Field: Exploring Whether Professionals Play Minimax in Laboratory Experiments," *Econometrica* 78 (2010): 1413–34.

27. P. E. Tetlock, *Expert Political Judgment: How Good Is It? How Can We Know?* (Princeton, NJ: Princeton University Press, 2005). Other researchers have found that experts' beliefs about their knowledge of a subject can actually lead to the creation of false memories, which leads to poor choices. Experts make up facts that don't exist to make their perception of the world conform to

their supposedly more knowledgeable view. See R. Mehta, J. Hoegg, and A. Chakravarti, "Knowing Too Much: Expertise-Induced False Recall Effects in Product Comparison," *Journal of Consumer Research* 38 (2011): 535–54. Experts and regulators are potentially prone to other types of biases: action bias, motivated reasoning, the focusing illusion, and the affect heuristic. See S. Tasic, "The Illusion of Regulatory Competence," *Critical Review* 21 (2009): 423–36.

28. R. Roberts, "Gambling with Other People's Money: How Perverted Incentives Caused the Financial Crisis," Mercatus Center Publication, George Mason University, April 28, 2010, mercatus.org/publication/gambling-other-peoples-money.

29. Brian Wansink, *Mindless Eating: Why We Eat More Than We Think* (New York: Bantam Dell, 2006), p. 11. If the recommendations in his book are not enough, then one can turn to his website and join Wansink's weight-loss plan—the Mindless Method (see www.mindlessmethod.com/).

30. R. H. Thaler and C. R. Sunstein, "Libertarian Paternalism," *American Economic Review* 93 (2003): 175.

31. Thaler and Sunstein later added this alternative to their list in *Nudge,* written five years later—perhaps in response to critiques by people such as Robert Sudgen (2008) (see note 32), who has made the same point I am making here. When later mentioning the alternative in their book, they dismiss the option by asking, "But should [the manager] Carolyn really try to maximize profits if the result is to make children less healthy, especially since she works for the school district?" The claim that paternalism is inevitable must be particularly narrow if it can be applied only in such special cases.

32. R. Sugden, "Why Incoherent Preferences Do Not Justify Paternalism," *Constitutional Politics and Economics* 19 (2008): 243.

33. R. Sugden, "Opportunity as a Space for Individuality: Its Value and the Impossibility of Measuring It," *Ethics* 113 (2003): 785.

34. A. K. Sen, *Inequality Reexamined* (Cambridge, MA: Harvard University Press, 1995), p. 41.

35. "The State Is Looking After You," *The Economist,* April 8, 2006.

36. Sugden, "Why Incoherent Preferences Do Not Justify Paternalism," p. 247.

CHAPTER 5: THE FASHION FOOD POLICE

1. The quote is by Urvashi Rangan, director of technical policy for the Consumers Union during a National Public Radio debate about the merits of organic food, held at New York University as a part of the program Intelligence Squared (see intelligencesquared us.org/wp-content/uploads/Organic-041310.pdf).

2. See the *Oprah Winfrey Show,* January 1, 2006, "Eat Organically," excerpts available online at www.oprah.com/world/Eat -Organically/1#ixzz1ZFMoGWcu.

3. W. D. McBride and C. Greene, "Characteristics, Costs, and Issues for Organic Dairy Farming," U.S. Department of Agriculture, Economic Research Service, Research Report Number 82, November 2009, www.ers.usda.gov/Publications/ERR82/ ERR82.pdf.

4. C. Greene et al., "New Directions in U.S. Organic Policy," in *Emerging Issues in the U.S. Organic Industry,* U.S. Department of Agriculture, Economic Research Service, Economic Information Bulletin No. EIB-55, June 2009, www.ers.usda.gov/publications/ eib55/eib55d.pdf. See also: www.ers.usda.gov/Briefing/Organic/ ProgramProvisions.htm.

5. www.oprah.com/world/Eat-Organically/1#ixzz1ZFSCsii4.

6. www.wholeliving.com/article/fresh-thinking-how-to-shop-for -fruits-and-vegetables?page=2.

7. For evidence on the fact that retail prices of organics are often more than 100 percent higher than conventional, see J. B. Chang, J. L. Lusk, and F. B. Norwood, "The Price of Happy Hens: A Hedonic Analysis of Retail Egg Prices," *Journal of Agricultural and Resource Economics* 35 (2010): 406–23; K. Brooks and J. L. Lusk, "Stated and Revealed Preferences for Organic and Cloned Milk: Combining Choice Experiment and Scanner Data," *American Journal of Agricultural Economics* 92 (2010): 1229–41; G. D. Thompson and J. Kidwell, "Explaining the Choice of Organic Produce: Cosmetic Defects, Prices, and Consumer

Preferences," *American Journal of Agricultural Economics* 80 (1998): 277–87. Data for more commodities are available at www .ers.usda.gov/Data/OrganicPrices/.

8. www.organic-market.info/web/News_in_brief/Food_Quality/ Italy/176/187/0/11501.html.

9. organic.about.com/od/marketingpromotion/tp/6-Largest -Organic-Retailers-In-North-America-2011.htm.

10. Examples of such studies include: L. Johansson et al., "Preference for Tomatoes, Affected by Sensory Attributes and Information about Growth Conditions," *Food Quality and Preference* 10 (1999): 289–98; Å Haglund et al., "Sensory Evaluation of Carrots from Ecological and Conventional Growing Systems," *Food Quality and Preference* 10 (1998): 23–29; A. Poelman et al., "The Influence of Information about Organic Production and Fair Trade on Preferences for and Perception of Pineapple," *Food Quality and Preference* 19 (2008): 114–21; L. Fillion and S. Arazi, "Does Organic Food Taste Better? A Claim Substantiation Approach," *Nutrition and Food Science* 32 (2002): 153–57; D. Bourn and J. John Prescott, "A Comparison of the Nutritional Value, Sensory Qualities, and Food Safety of Organically and Conventionally Produced Foods," *Critical Reviews in Food Science and Nutrition* 42 (2002): 1–34; A. E. Croissant et al., "Chemical Properties and Consumer Perception of Fluid Milk from Conventional and Pasture-Based Production Systems," *Journal of Dairy Science* 90 (2007): 4942–53; A. Jonsäll et al., "Effects of Genotype and Rearing System on Sensory Characteristics and Preference for Pork (*M. Longissimus dorsi*)," *Food Quality and Preference* 13 (2002): 73–80.

11. Brian Wansink, *Mindless Eating: Why We Eat More Than We Think* (New York: Bantam Dell, 2006).

12. S. M. McClure et al., "Neural Correlates of Behavioral Preference for Culturally Familiar Drinks," *Neuron* 22 (2004): 379–87.

13. Napolitano et al., "Effect of Information About Organic Production on Beef Liking and Consumer Willingness to Pay," *Food Quality and Preference* 21 (2010): 207–12.

14. C. Baumrucker, "Why Does Organic Milk Last So Much Longer

Than Regular Milk?" ScientificAmerican.com: Ask the Experts, June 6, 2008, available online at www.scientificamerican.com/article.cfm?id=experts-organic-milk-lasts-longer.

15. Don't take my word for it. See Clare et al., "Comparison of Sensory, Microbiological, and Biochemical Parameters of Microwave Versus Indirect UHT Fluid Skim Milk During Storage," *Journal of Dairy Science* 88 (2005): 4172–82.

16. Dangour et al., "Nutritional Quality of Organic Foods: A Systematic Review," *American Journal of Clinical Nutrition* (September 2009).

17. Quote is reported in J. P. Schuldt and N. Schwarz, "The "Organic" Path to Obesity? Organic Claims Influence Calorie Judgments and Exercise Recommendations," *Judgment and Decision Making* 5 (2010): 144–50.

18. Ibid.

19. See Bjørn Lomborg, *The Skeptical Environmentalist* (New York: Cambridge University Press, 2001), p. 233, and the references therein.

20. C. K. Winter and S. F. David, "Organic Foods," *Journal of Food Science* 71 (2006): E117–R124.

21. Lomborg, *The Skeptical Environmentalist*, p. 245. See also page 234 for a comparison of the cancer risk from different foods.

22. Ibid., p. 247. Lomborg also presents the numbers of deaths from radon used in the following paragraph.

23. www.foodsafetynews.com/2011/06/german-e-coli-crisis-refuels-irradiation-debate/.

24. Of course, there were recalls of nonorganic foods, too, during this time. It is also true that many more nonorganic foods are recalled each year than organic. But you have to keep in mind that many more nonorganic foods are produced each year. Only about 0.7 percent of all U.S. cropland and 0.5 percent of all U.S. pasture was certified organic in 2008 (www.ers.usda.gov/data/organic/). So, the real question is the rate of food recalls *per pound* of food produced. As far as I know, no such statistic has been calculated.

25. Mukherjee et al., "Preharvest Evaluation of Coliforms, Escherichia coli, Salmonella, and Escherichia coli O157:H7 in Organic and Conventional Produce Grown by Minnesota Farmers," *Journal of Food Protection* 67 (2004): 894–900.

26. J. Horowitz et al., *"No-Till" Farming Is a Growing Practice*, U.S. Department of Agriculture, Economic Research Service, Economic Information Bulletin Number 70, November 2010, www.ers.usda.gov/Publications/EIB70/EIB70.pdf.

27. www.dasnr.okstate.edu/notill/hidden/nt_ok_2009_pres/NT_OK _SURVEY.pdf.

28. Mader et al., "Soil Fertility and Biodiversity in Organic Farming," *Science* 296 (2002): 1694–97.

29. McBride and Greene, "Characteristics, Costs, and Issues for Organic Dairy Farming."

30. U.S. Department of Agriculture, Census of Agriculture, *Organic Production Survey* 3, Special Studies Part 2, AC-07-SS-2 (2008), issued February 2010, updated July 2010 www.agcensus.usda.gov/ Publications/2007/Online_Highlights/Organics/ORGANICS .pdf; L. Gianessi and A. Rinkus, "USDA 2008 National Organic Production Survey: National Acreage and Crop Yields," white paper, CropLife Foundation, March 2010, www.croplife foundation.org/Organics/USDA%202008%20National%20 Organic%20Production%20Survey%20Analysis.pdf.

31. www.ers.usda.gov/data/organic/.

32. M. Stewart, "Food Is the New Fashion," *HuffingtonPost.com*, February 9, 2011, www.huffingtonpost.com/martha-stewart/food -is-the-new-fashion_1_b_821078.html.

33. Ibid.

34. L. Miller, "Divided We Eat," *Newsweek*, November 29, 2010.

35. bittman.blogs.nytimes.com/2011/09/24/is-junk-food-really -cheaper/.

36. S. Clifford, "The Plastic Sandwich Bag Flunks," *New York Times*, August 26, 2011, p. B1 of the New York edition.

37. See J. L. Lusk and F. B. Norwood, "Bridging the Gap between Laboratory Experiments and Naturally Occurring Markets: An Inferred Valuation Method," *Journal of Environmental Economics and Management* 58 (2009): 236–50, and J-B. Chang, J. L. Lusk, and F. B. Norwood, "How Closely Do Hypothetical Surveys and Laboratory Experiments Predict Field Behavior?" *American Journal of Agricultural Economics* 91 (2009): 518–34.

CHAPTER 6: FRANKEN-FEARS

1. D. Q. Fuller and L. Qin, "Water Management and Labour in the Origins and Dispersal of Asian Rice," *World Archaeology* 41 (2009): 88–111.

2. K. F. Kiple and K. C. Ornelas, eds., "II.B.3.—Potatoes (White)," *The Cambridge World History of Food* (Cambridge, UK: Cambridge University Press, 2000).

3. www.telegraph.co.uk/earth/earthnews/3349308/Prince-Charles -warns-GM-crops-risk-causing-the-biggest-ever-environmental -disaster.html.

4. www.ers.usda.gov/Data/BiotechCrops/.

5. J. Fernandez-Cornejo and M. Caswell, *The First Decade of Genetically Engineered Crops in the United States,* U.S. Department of Agriculture, Economic Research Service, Economic Information Bulletin No. 11, April 2006, p. 11, www.ers.usda.gov/ publications/eib11/eib11.pdf.

6. M. Qaim and D. Zilberman, "Yield Effects of Genetically Modified Crops in Developing Countries," *Science* 299 (2003): 900–902.

7. www8.nationalacademies.org/onpinews/newsitem.aspx ?RecordID=12804.

8. "Safety of Genetically Engineered Foods: Approaches to Assessing Unintended Health Effects," National Academies Press (Washington, DC: National Academies Press, 2004), www.nap.edu/catalog.php?record_id=10977; American Dietetic Association (ADA), "Position of the American Dietetic Association: Agricultural and Food Biotechnology," *Journal of the American Dietetic Association,* 106, no. 2 (2006): 285–93; American Medical Association, Recommendations of Council on Scientific Affairs Reports, AMA Interim Meeting, 2000, http:// www.ama-assn.org/resources/doc/csaph/csai-00.pdf.; United Nations, Food and Agriculture Organization (FA), position statement on biotechnology, http://www.fao.org/biotech/fao -statement-on-biotechnology/en/; FAO/World Health Organization, "Safety Aspects of Genetically Modified Foods of

Plant Origin: A Joint FAO/WHO Consultation on Foods Derived from Biotechnology," Geneva, Switzerland, May 29–June 2, 2000, www.who.int/foodsafety/publications/biotech/ec_june2000/en/index.html.

9. J. B. Falck-Zepeda, G. Traxler, and R. G. Nelson, "Surplus Distribution from the Introduction of a Biotechnology Innovation," *American Journal of Agricultural Economics* 82 (2000): 360–69.

10. G. K. Price, W. Lin, J. B. Falck-Zepeda, and J. Fernandez-Cornejo, "The Size and Distribution of Market Benefits from Adopting Agricultural Biotechnology," U.S. Department of Agriculture, Economic Research Service, Technical Bulletin No. 1906, November 2003.

11. M. Qaim, "The Economics of Genetically Modified Crops," *Annual Review of Resource Economics* 1 (2009): 665–93.

12. See table 3 in Jorge Fernandez-Cornejo, "The First Decade of Genetically Engineered Crops in the United States," U.S. Department of Agriculture, Economic Research Service, Economic Information Bulletin 11, April 2006.www.ers.usda.gov/publications/eib11/eib11.pdf.

13. R. Paarlberg, *Starved for Science: How Biotechnology Is Being Kept Out of Africa* (Cambridge, MA: Harvard University Press, 2008).

14. Quoted in J. Tierney, "Greens and Hunger, TierneyLab: Putting Ideas in Science to the Test," *New York Times*, May 19, 2008, tierneylab.blogs.nytimes.com/2008/05/19/greens-and-hunger/?pagemode=print.

CHAPTER 7: THE FOLLIES OF FARM POLICY

1. M. Pollan, "Farmer in Chief," *New York Times*, October 9, 2008. A shorter version appeared on October 12, 2008, on page MM62 of the New York edition, www.nytimes.com/2008/10/12/magazine/12policy-t.html?pagewanted=all.

2. M. Bittman, "Don't End Agricultural Subsidies. Fix Them," *New York Times*, March 2, 2011, query.nytimes.com/gst/fullpage.html?res=9A0CEFD8133EF931A35750C0A9679D8B63&scp=8&sq=mark+bittman&st=nyt.

3. Pollan, "Farmer in Chief."

4. M. Bittman, "Bad Food? Tax It, and Subsidize Vegetables," *New York Times*, July 23, 2011, www.nytimes.com/2011/07/24/opinion/sunday/24bittman.html?pagewanted=2&_r=2&sq=mark bittman&st=nyt&scp=5.

5. www.thenation.com/article/163401/resisting-corporate-theft-seeds.

6. Data come from the table "Outlays," on page 121 of *FY2012 Budget Summary and Annual Performance Plan*, U.S. Department of Agriculture, www.obpa.usda.gov/budsum/FY12budsum.pdf.

7. B. K. Goodwin and A. K. Mishra, "Are 'Decoupled' Farm Program Payments Really Decoupled? An Empirical Evaluation," *American Journal of Agricultural Economics* 88 (2006): 73–89.

8. W. Lin and R. Dismukes, "Supply Response Under Risk: Implications for Counter-Cyclical Payments' Production Impact," *Applied Economic Perspectives and Policy* 28 (2007): 64–86.

9. A. M. Okrent and J. M. Alston, "The Effects of Farm Commodity and Retail Food Policies on Obesity and Economic Welfare in the United States," *American Journal of Agricultural Economics* 94 (2012) 611–46.

10. J. M. Alston, "Benefits and Beneficiaries from U.S. Farm Subsidies," AEI Agricultural Policy Series: The 2007 Farm Bill and Beyond, American Enterprise Institute, 2007; D. McDonald et al., *U.S. Agriculture Without Farm Support*, Research Report 06.10, ABARE (Australian Bureau of Agricultural and Resource Economics), Canberra (September 2006).

11. www.fao.org/news/story/en/item/20568/icode/.

12. For example, see N. Z. Muller, R. Mendelsohn, and W. Nordhaus, "Environmental Accounting for Pollution in the United States Economy," *American Economic Review* 101 (2011): 1649–75.

13. Michael Pollan, *The Omnivore's Dilemma: A Natural History of Four Meals* (New York: Penguin, 2006), p. 64.

14. M. Duffy and A. Holste, "Estimated Returns to Iowa Farmland," *Journal of the American Society of Farm Managers and Rural Appraisers* 68 (2005): 102–109.

15. Julian M. Alston, "The Incidence of U.S. Farm Programs," *The Economic Impact of Public Support to Agriculture*, V. E. Ball et al., eds., vol. 7, pt. 1 (New York: Springer, 2010), pp. 81–105.

16. B. E. Kirwan, "The Distribution of U.S. Agricultural Subsidies," The American Enterprise Institute Agricultural Policy Series, 2007 Farm Bill and Beyond: Summary for Policymakers, Bruce L. Gardner and Daniel A. Sumner, eds., 2007.

17. Pollan, *The Omnivore's Dilemma*, p. 220.

18. www.ers.usda.gov/briefing/wellbeing/demographics.htm.

19. R. A. Hoppe et al., "Structure and Finances of U.S. Farms: Family Farm Report," USDA-ERS, Economic Information Bulletin No. EIB-24, June 2007.

20. Ibid.

21. www.ers.usda.gov/Briefing/WellBeing/farmhouseincome.htm.

22. See, for example, www.ers.usda.gov/briefing/farmincome/ govtpaybyfarmtype.htm. In this sentence, "small farms" refers to those with less than $50,000 in annual farm sales; large farms have more than $500,000 in annual farm sales.

23. Kirwan, "The Distribution of U.S. Agricultural Subsidies."

24. www.ers.usda.gov/amberwaves/november08/Findings/ NewPayment.htm.

25. In 2010 the median income of farm households in the United States was over $54,370, increasing 4.1 percent from 2009. By contrast, the median household income of all households in the United States was only $49,445 in 2010, a 2.3 percent *decrease* from 2009, www.census.gov/newsroom/releases/archives/income _wealth/cb11-157.html.

26. Public Interest Research Group, "Apples to Twinkies: Comparing Federal Subsidies of Fresh Produce and Junk Food," Report released on Wednesday, September 21, 2011. www.uspirg.org/ reports/xxp/apples-twinkies.

27. J. M. Alston, D. A. Sumner, and S. A. Vosti, "Farm Subsidies and Obesity in the United States: National Evidence and International Comparisons," *Food Policy* 33 (2008): 470–79.

28. J. C. Miller and K. H. Coble, "Cheap Food Policy: Fact or Rhetoric?" *Food Policy* 32 (2007): 98–111; D. Cutler, E. Glaeser, and J. Shapiro, "Why Have Americans Become More Obese?" *Journal of Economic Perspectives* 17 (2003): 93–118.

29. Alston, "Benefits and Beneficiaries from U.S. Farm Subsidies."

30. J. C. Beghin and H. H. Jensen, "Farm Policies and Added Sugars in US Diets," *Food Policy* 33 (2008): 480–88.

31. www.ers.usda.gov/data/farmtoconsumer/pricespreads.htm.

32. Data come from the table "Outlays" on p. 121 of "FY2012 Budget Summary and Annual Performance Plan," www.obpa.usda.gov/budsum/FY12budsum.pdf.

33. See data in www.census.gov/compendia/statab/2012/tables/12s0844.pdf.

34. Bittman, "Don't End Agricultural Subsidies. Fix Them."

35. These data are based on the author's calculations using the 2002 National Pesticide Use Database, at www.croplifefoundation.org/cpri_npud2002.htm.

36. The statistics are calculated using the midpoints of the water use ranges reported in table 14, "Indicative Values of Crop Water Needs and Sensitivity to Drought," in FAO, *Irrigation Water Management: Irrigation Water Needs,* www.fao.org/docrep/S2022E/s2022e07.htm. For reference, corn requires 13 percent more water than soybeans and wheat requires 4 percent less water than soybeans.

37. www.goodfruit.com/Good-Fruit-Grower/January-1st-2011/Specialty-crops-will-be-on-center-stage/.

38. B. Wolfgang, "'Healthier' School Lunch at What Cost?" *Washington Times,* May 16, 2011, www.washingtontimes.com/news/2011/may/16/healthier-school-lunch-at-what-cost/?page=all.

39. F. Santos, "Public Schools Face the Rising Costs of Serving Lunch," *New York Times,* September 19, 2011, www.nytimes.com/2011/09/20/education/20lunch.html?_r=1.

40. See J. A. Kropski et al., "School-based Obesity Prevention Programs: An Evidence-based Review," *Obesity* 16 (2008): 1009–18.

41. A. Ishdorj, H. H. Jensen, and M. K. Crepinsek, "Children's Consumption of Fruits and Vegetables: School Meals vs. Home Meals," paper presented at the Agricultural and Applied Economics Association meeting in Denver, CO, July 27, 2010.

42. D. R. Taber et al., "Banning All Sugar-Sweetened Beverages in Middle Schools," *Archives of Pediatrics and Adolescent Medicine,* published online November 7, 2011, doi:10.1001/archpediatrics .2011.200; J. M. Fletcher, D. Frisvold, and N. Tefft, "Taxing Soft Drinks and Restricting Access to Vending Machines to Curb Child Obesity," *Health Affairs* 29 (2010): 1059–66.

43. K. Ralston et al., "The National School Lunch Program: Background, Trends, and Issues," U.S. Department of Agriculture, Economic Research Service, Economic Research Report No. 61, July 2008, www.ers.usda.gov/publications/err61/ err61.pdf.

44. C. Kimball, "Letter from Vermont," America's Test Kitchen, January 25, 2012, www.americastestkitchenfeed.com/notes-from -cpk/2012/01/january-letter-from-vermont/.

45. Pollan, "Farmer in Chief."

CHAPTER 8: THE THIN LOGIC OF FAT TAXES

1. www.cdc.gov/cdctv/ObesityEpidemic/.

2. D. L. Costa and R. H. Steckel, "Long-Term Trends in Health, Welfare, and Economic Growth in the United States," in R. H. Steckel and R. Floud, eds., *Health and Welfare During Industrialization* (Chicago, IL: University of Chicago Press, 1997), pp. 47–89.

3. J. E. Oliver, *Fat Politics* (New York: Oxford University Press, 2006), p. 7.

4. The relevant papers are D. B. Allison et al., "Annual Deaths Attributable to Obesity in the United States," *Journal of the American Medical Association* 282 (1999): 1530–38; A. H. Mokdad et al., "Actual Causes of Death in the United States," *Journal of the American Medical Association* 291 (2004): 1238–45; A. H. Mokdad et al., "Correction: Actual Causes of Death in the United States, 2000," *Journal of the American Medical Association* 293 (2005): 293–94; K. Flegal et al., "Excess Deaths Associated with Underweight, Overweight, and Obesity," *Journal of the American Medical Association* 293 (2005): 1861–67.

5. www.cdc.gov/NCHS/data/nvsr/nvsr58/nvsr58_19.pdf.

6. For accounts of these developments, see P. Campos *The Obesity Myth* (New York: Gotham, 2004), or Oliver, *Fat Politics*. On the change in overweight definitions, see www.washingtonpost.com/wp-srv/style/guideposts/fitness/optimal.htm.

7. These estimates are inferred from table 2 in K. Flegal et al., "Excess Deaths Associated with Underweight, Overweight, and Obesity."

8. Oliver, *Fat Politics*, p. 4.

9. E. W. Gregg et al., "Secular Trends in Cardiovascular Disease Risk Factors According to Body Mass Index in US Adults," *Journal of the American Medical Association* 293 (2005): 1868–74.

10. www.cdc.gov/diabetes/pubs/pdf/ndfs_2011.pdf.

11. A. H. Mokdad et al., "Prevalence of Obesity, Diabetes, and Obesity-Related Health Risk Factors, 2001," *Journal of the American Medical Association* 289 (2003): 76–79.

12. www.diabetes.org/diabetes-basics/diabetes-myths/.

13. Annual average changes in prevalence rates were taken from www.cdc.gov/diabetes/statistics/prev/national/figage.htm.

14. C. L. Ogden et al., "Prevalence and Trends in Overweight Among U.S. Children and Adolescents, 1999–2000," *Journal of the American Medical Association* 288 (2002): 1728–32.

15. www.cdc.gov/diabetes/projects/cda2.htm.

16. dailycaller.com/2012/01/25/nyc-food-police-caught-in-an-ad-lie/.

17. Y. Li et al., "Declining Rates of Hospitalization for Nontraumatic Lower-Extremity Amputation in the Diabetic Population Aged 40 Years or Older: U.S., 1988–2008," *Diabetes Care* 35 (2012): 273–77.

18. www.cdc.gov/nchs/data/hus/hus10.pdf#022.

19. E. I. Lubetkin and H. Jia, "Health-Related Quality of Life, Quality-Adjusted Life Years, and Quality-Adjusted Life Expectancy in New York City from 1995 to 2006," *Journal of Urban Health* 6 (2009): 551–61.

20. K. M. Flegal et al., "Prevalence and Trends in Obesity Among US Adults, 1999–2008," *Journal of the American Medical Association* 303 (2010): 235–41.

21. www.ibtimes.com/articles/177222/20110709/obesity-rates-study
    -rise-health-robert.htm.

22. Data are in K. M. Flegal et al., "Overweight and Obesity in the
    United States: Prevalence and Trends, 1960–1994," *International
    Journal of Obesity and Related Metabolic Disorders* 22 (1998):
    39–47; C. L. Ogden et al., "Prevalence of Overweight and Obesity
    in the United States, 1999–2004," *Journal of the American Medical
    Association* 295 (2006): 1549–55; K. M. Flegal et al., "Prevalence
    and Trends in Obesity Among US Adults, 1999–2008."

23. First quote is by Bruce Silverglade from CSPI in Brussels at
    "Generating Excess: A Conference on Diet, Obesity, and Public
    Policy," February 3, 2004. The second quote is by Margo Wootan
    at CSPI, www.foxnews.com/story/0,2933,112546,00.html.

24. E. Allday, "UCSF Scientists Declare War on Sugar in Food," *San
    Francisco Chronicle,* February 2, 2012, www.sfgate.com/cgi-bin/
    article.cgi?f=/c/a/2012/02/01/BA891N1PQS.DTL.

25. Thomas Sowell, *The Vision of the Anointed: Self-Congratulation as
    a Basis for Social Policy* (New York: Basic Books, 1996), p. 203.

26. See J. Bhattacharya and N. Sood, "Who Pays for Obesity?" *Journal
    of Economic Perspectives* 25 (2011): 139–58.

27. M. Bittman, "Bad Food? Tax It, and Subsidize Vegetables," *New
    York Times,* July 24, 2011, www.nytimes.com/2011/07/24/opinion/
    sunday/24bittman.html?scp=5&sq=mark+bittman&st=nyt.

28. Barry Popkin, *The World Is Fat* (New York: Avery/Penguin USA,
    2008), pp. 162–63.

29. K. Brownell, "Get Slim with Higher Taxes," *New York Times,*
    December 15, 1994.

30. www.impacteen.org/obesitystatedata.htm#01.

31. Kuchler et al., "Taxing Snack Foods: Manipulating Diet Quality
    or Financing Information Programs?" *Review of Agricultural
    Economics* 27 (2005): 4–20.

32. H. H. Chouinard et al., "Fat Taxes: Big Money for Small Change,"
    Forum for Health Economics and Policy 10 (2007): article 2.

33. S. Dharmasena and O. Capps Jr., forthcoming, "Intended and
    Unintended Consequences of a Proposed National Tax on
    Sugar-Sweetened Beverages to Combat the U.S. Obesity Problem,"

Health Economics, forthcoming, onlinelibrary.wiley.com/
doi/10.1002/hec.1738/abstract; J. L. Lusk and C. Schroeter, "When
Do Fat Taxes Increase Consumer Welfare?" *Health Economics*,
forthcoming, onlinelibrary.wiley.com/doi/10.1002/hec.1789/
abstract?userIsAuthenticated=false&deniedAccessCustomised
Message=.

34. R. Sturm et al., "Soda Taxes, Soft Drink Consumption, and
    Children's Body Mass Index," *Health Affairs* 29 (2010): 1052–58.

35. C. Schroeter, J. Lusk, and W. Tyner, "Determining the Impact of
    Food Price and Income Changes on Body Weight," *Journal of
    Health Economics* 27 (2008): 45–68.

36. H. H. Chouinard et al., "Fat Taxes."

37. Bittman, "Bad Food?"

38. http://www.oneidadispatch.com/articles/2011/11/27/news/
    doc4ed2f432b3335533265968.txt.

39. "Tobacco Settlement: States' Allocations of Fiscal Year 2003 and
    Expected Fiscal Year 2004 Payments," U.S. General Accounting
    Office (GAO), March 22, 2004.

40. M. F. Teisl, N. E. Bockstael, and A. Levy, "Measuring the Welfare
    Effects of Nutrition Information," *American Journal of
    Agricultural Economics* 83 (2001): 133–49; K. Kiesel and S. B.
    Villas-Boas, "Can Information Costs Affect Consumer Choice?
    Nutritional Labels in a Supermarket Experiment," *International
    Journal of Industrial Organization*, in press; A. C. Drichoutis,
    R. M. Nayga Jr., and P. Lazaridis, "Can Nutritional Label
    Use Influence Body Weight Outcomes?" *Kyklos* 62 (2009):
    500–25.

41. J. S. Downs, G. Loewenstein, and J. Wisdom, "Strategies for
    Promoting Healthier Food Choices," *American Economic Review*
    99 (2009): 1–10; J. Wisdom, J. S. Downs, and G. Loewenstein,
    "Promoting Healthy Choices: Information versus Convenience,"
    American Economic Journal: Applied Economics 2 (2010): 164–78;
    B. Bollinger, P. Leslie, and A. Sorensen, "Calorie Posting in Chain
    Restaurants," *American Economic Journal: Economic Policy* 3
    (2011): 91–128; B. Ellison, D. Davis, and J. L. Lusk, "The Value

and Cost of Restaurant Calorie Labels: Results from a Field Experiment," working paper, Oklahoma State University.

42. J. S. Downs, G. Loewenstein, and J. Wisdom, "Eating by the Numbers," *New York Times,* op-ed, November 13, 2009.

43. www.freakonomics.com/2011/08/26/mandating-calorie-counts -has-libertarian-paternalism-gone-too-far/.

44. Popkin, *The World Is Fat,* p. 165.

45. P. M. Ippolito and A. D. Mathios, "Information, Advertising and Health Choices: A Study of the Cereal Market," *The RAND Journal of Economics* 21 (1990): 459–80.

46. www.rollcall.com/issues/57_32/lee_terry_paul_broun _nutritional_guidelines_nanny_state_run_amok-208873-1 .html?pos=oopih.

47. www2.tbo.com/news/opinion/2011/oct/17/meopino1-junk-food -government-ar-272223/.

48. CSPI's Nutrition Action Healthletter, July/August 1998, U.S. edition, www.cspinet.org/nah/7_98eat.htm.

49. M. Nestle, *Food Politics* (Berkeley, CA: University of California Press, 2007).

50. J. Sullum, "War on Fat: Is the Size of Your Butt the Government's Business?" *Reason,* August/September 2004. reason.com/ archives/2004/08/01/the-war-on-fat/singlepage.

51. L. Murtagh and D. S. Ludwig, "State Intervention in Life-Threatening Childhood Obesity," *Journal of the American Medical Association* 306 (2011): 206–7.

52. F. Etilé, "Changes in the Distribution of the Body Mass Index in France, 1981–2003: A Decomposition Analysis," INRA ALISS Working Paper 2011-02 (April 2011).

53. C. L. Baum and S. Y. Chou, "The Socio-Economic Causes of Obesity," NBER Working Paper No. 17423, September 2011.

54. D. M. Cutler, E. L. Glaeser, and J. M. Shapiro, "Why Have Americans Become More Obese?" *Journal of Economic Perspectives* 17 (2003): 93–118.

55. J. Sullum, "War on Fat: Is the Size of Your Butt the Government's Business?"

CHAPTER 9: THE LOCAVORE'S DILEMMA

1. D. Brown, "Caterers Find Eco-Standards Tough to Chew," *Denver Post,* May 19, 2008, www.denverpost.com/lifestyles/ci_9305736.

2. *Time,* March 13, 2007.

3. M. Pollan, "Farmer in Chief," *New York Times,* October 9, 2008.

4. Kim Severson, *Spoon Fed* (New York: Riverhead Books/Penguin USA, 2010), pp. 67–68.

5. Ibid., p. 69.

6. J. E. McWilliams, *Just Food* (New York: Back Bay Books/Hachette, 2009), p. 30.

7. *Bill Moyers Journal,* November 28, 2008, interview with Michael Pollan, at www.pbs.org/moyers/journal/11282008/transcript1 .html.

8. S. Nasar, *Grand Pursuit: The Story of Economic Genius* (New York: Simon & Schuster, 2011), p. 392.

9. For changes in agricultural productivity and labor usage, see www.ers.usda.gov/publications/EB9/eb9.pdf.

10. See data in table 7 of E. K. Mafoua and F. Hossain, "Scope and Scale Economies for Multi-Product Farms: Firm-Level Panel Data Analysis," selected paper presented at the American Agricultural Economics Association Meeting, Chicago, August 5–8, 2001.

11. Figure comes from author's calculation using the fourth column of results in table 7 in R. Mosheim and C. A. Knox Lovell, "Scale Economies and Inefficiency of U.S. Dairy Farms," *American Journal of Agricultural Economics* 91 (2009): 777–94.

12. V. Postrel, "No Free Locavore Lunch," *Wall Street Journal,* September 25, 2010, online.wsj.com/article/SB10001424052748703 989304575503972940124574.html; for evidence on the higher prices of local, see J. Biermacher et al., "Economic Challenges of Small-Scale Vegetable Production and Retailing in Rural Communities: An Example from Rural Oklahoma," *Journal of Food Distribution Research* 37 (2007): 1–13.

13. For just one example of the claim, see J. Rickardson, "5 Ridiculous Myths People Use to Trash Local Food—and Why They're Wrong," Alternet.com, November 18, 2011, www.alternet

.org/food/153121/5_ridiculous_myths_people_use_to_trash
_local_food_—_and_why_they're_wrong?page=entir.

14. Michael Pollan, *The Omnivore's Dilemma: A Natural History of Four Meals* (New York: Penguin, 2006), p. 222.

15. G. Edwards-Jones et al., "Testing the Assertion That 'Local Food Is Best': The Challenges of an Evidence Based Approach," *Trends in Food Science and Technology* 19 (2008): 265–74.

16. C. Saunders, A. Barber, and G. Taylor, "Food Miles— Comparative Energy/Emissions Performance of New Zealand's Agriculture Industry," Lincoln University, Agribusiness and Economics Research Unit, Research report no. 285, July 2006.

17. See M. F. Teisl, "Environmental Concerns in Food Consumption," in *Oxford Handbook of the Economics of Food Consumption and Policy*, J. L. Lusk, J. Roosen, and J. F. Shogren, eds. (Oxford: Oxford University Press, 2011), and the references therein.

18. Ibid., p. 850.

19. E. Engelhaupt, "Do Food Miles Matter?" *Environmental Science and Technology* 42 (2008): 3482.

20. D. Coley, M. Howard, and M. Winter, "Local Food, Food Miles and Carbon Emissions: A Comparison of Farm Shop and Mass Distribution Approaches," *Food Policy* 34 (2009): 150–55.

21. articles.boston.com/2011-06-16/bostonglobe/29666344_1 _greenhouse-gas-carbon-emissions-local-food.

22. Biermacher et al., "Economic Challenges of Small-Scale Vegetable Production and Retailing in Rural Communities."

23. M. Bittman, "Local Food: No Elitist Plot," nytimes.com, November 1, 2011, opinionator.blogs.nytimes.com/2011/11/01/ local-food-no-elitist-plot/.

24. C. Price, "The Locavore's Dilemma: What to Do with the Kale, Turnips, and Parsley That Overwhelm Your CSA Bin," Slate.com, June 8, 2010, www.slate.com/articles/news_and_politics/ recycled/2010/06/the_locavores_dilemma.html.

25. J. C. Rickman et al., "Nutritional Comparison of Fresh, Frozen and Canned Fruits and Vegetables, Part 1. Vitamins C and B and Phenolic Compounds," *Journal of the Science of Food and Agriculture* 87 (2007): 930–44; J. C. Rickman et al., "Nutritional

Comparison of Fresh, Frozen, and Canned Fruits and Vegetables, Part 2. Vitamin A and Carotenoids, Vitamin E, Minerals and Fiber," *Journal of the Science of Food and Agriculture* 87 (2007): 1185–96; D. J. Favell, "A Comparison of the Vitamin C Content of Fresh and Frozen Vegetables," *Food Chemistry* 62 (1998): 59–64.

26. L. S. Drescher et al., "A New Index to Measure Healthy Food Diversity Better Reflects a Healthy Diet Than Traditional Measures," *Journal of Nutrition* 137 (2007): 647–51.

27. E.g., see www.peachstand.com/.

28. Severson, *Spoon Fed*, p. 75.

29. Pollan, *The Omnivore's Dilemma*, p. 252.

30. Ibid., p. 253.

31. Ibid., p. 260.

32. J. Rickardson, "5 Ridiculous Myths People Use to Trash Local Food—and Why They're Wrong."

33. www.nytimes.com/2011/08/30/science/30tierney.html ?pagewanted=all.

CHAPTER 10: THE FUTURE OF FOOD

1. A. C. Heller, *Ayn Rand and the World She Made* (New York: Nan A. Talese/Doubleday, 2009), p. 395.

2. Charles Mackay, *Extraordinary Popular Delusions and the Madness of Crowds*, 1841.

3. *Bourgeois Dignity: Why Economics Can't Explain the Modern World* (Chicago: University of Chicago Press, 2010).

4. Controlling population growth sounds like a good idea when it involves other people somewhere across the globe, but it doesn't sound so good when it moves closer to home. Who, after all, gets to decide the number of people who inhabit Earth? The reality is that population is endogenously determined with food prices and income, and—in a sense—it is somewhat self-regulating. Many of the arguments surrounding population control are eerily similar to those surrounding the eugenics movement of a century ago.

5. Walter Williams, *Liberty versus the Tyranny of Socialism: Controversial Essays* (Stanford, CA: Hoover Press, 2008), p. 78.

INDEX

Index

JAYSON LUSK is the most prolific academic food economist of the past decade, publishing more than one hundred articles in peer-reviewed scientific journals on topics related to consumer behavior and food marketing and policy. After serving on the faculty at Mississippi State University and Purdue University, and a stint as visiting researcher at the French National Institute for Agricultural Research, he is currently a professor and the Willard Sparks Endowed Chair in the agricultural economics department at Oklahoma State University. Lusk earned a Ph.D. in agricultural economics from Kansas State University in 2000 and a B.S. in food technology from Texas Tech University in 1997, where he was named Outstanding Student in the College of Agricultural and Life Sciences.

Lusk has served on the editorial councils of seven top academic journals, including the *American Journal of Agricultural Economics* and the *Journal of Environmental Economics and Management*. He has served on the executive committees of the three largest U.S. agricultural economics associations, won numerous research awards, and given lectures at more than thirty universities in the United States and abroad.

In 2007, Lusk coauthored a book on consumer research methods published by Cambridge University Press and coauthored an undergraduate textbook on agricultural marketing and price analysis. In 2011, he released a coauthored book on the economics of farm animal welfare published by Oxford University Press and he coedited the *Oxford Handbook on the Economics of Food Consumption and Policy*.

Jayson lives in Stillwater, Oklahoma, with his wife and their two sons.